Information Circular 9526

Pillar and Roof Span Design Guidelines for Underground Stone Mines

Gabriel S. Esterhuizen, Dennis R. Dolinar, John L. Ellenberger, and Leonard J. Prosser

DEPARTMENT OF HEALTH AND HUMAN SERVICES
Centers for Disease Control and Prevention
National Institute for Occupational Safety and Health
Office of Mine Safety and Health Research
Pittsburgh, PA • Spokane, WA

May 2011

This document is in the public domain and may be freely copied or reprinted.

Disclaimer

Mention of any company or product does not constitute endorsement by the National Institute for Occupational Safety and Health (NIOSH). In addition, citations to Web sites external to NIOSH do not constitute NIOSH endorsement of the sponsoring organizations or their programs or products. Furthermore, NIOSH is not responsible for the content of these Web sites.

Ordering Information

To receive documents or other information about occupational safety and health topics, contact NIOSH at

> Telephone: **1–800–CDC–INFO** (1–800–232–4636)
> TTY: 1–888–232–6348
> e-mail: cdcinfo@cdc.gov
>
> or visit the NIOSH Web site at **www.cdc.gov/niosh**.

For a monthly update on news at NIOSH, subscribe to NIOSH *eNews* by visiting **www.cdc.gov/niosh/eNews**.

DHHS (NIOSH) Publication No. 2011–171

May 2011

SAFER • HEALTHIER • PEOPLE™

Contents

Background 1
    Research Methodology 1
    Applicability of Design Guidelines 2
Geotechnical Characteristics 2
    Geological Setting 2
    Intact Rock Strength 3
    Rock Mass Characteristics 3
    Large Joints 4
    Rock Mass Rating 4
    Horizontal Rock Stress 5
Pillar Design Considerations 6
    Background 6
    Pillar Strength 6
    Pillar Stress Calculation 7
    Pillar Failure 7
    Safety Factor 8
    Developing a Pillar Strength Equation for Stone Mines 8
Survey of Stone Mine Pillar Performance 9
    Successful Pillar Layouts 9
    Failed Pillars 9
    Observed Rib Instability 13
    Wide-Area Failures 15
    Summary Chart of Pillar Observations 16
Stone Mine Pillar Stability Analysis 17
    Brittle Rock Spalling 17
    The Impact of Large, Angular Discontinuities 20
    The Impact of Weak Bedding Bands 23
    The Effect of Floor Benching 26
    Pillar Length Effect on Pillar Strength 32
Pillar Strength Equation for Stone Mines 33
    Base Equation 33
    Adjustment for the Presence of Large Discontinuities 34
    Adjustment for Rectangular Pillars 36
    Adjustments for Other Geotechnical Conditions 37
    Pillar Strength Equation Modified for Stone Mines 37
    Pillar Factor of Safety Determination 37
    Applicability of the Pillar Design Equation 39
Roof Span Design Considerations 39
    Background 39
    Methods of Roof Span Design 39
    Stability of Bedded Rock 40
    Developing Roof Span Design Guidelines for Stone Mines 40
Survey of Roof Span Performance 41
    Roof Instabilities 42
    Support Practices 44
    Comparison of Roof Stability in the Physiographic Regions 44

- Stone Mine Roof Stability Analysis 45
    - Roof Span Dimensions 45
    - Thickness of the Immediate Roof Beam 47
    - Horizontal Stress Issues 48
    - Roof Support 54
- Pillar and Roof Span Design Guidelines 54
    - Geotechnical Characterization 55
    - Roof Span Selection 55
    - Selecting the Roof Horizon 56
    - Orientation of Headings 56
    - Roof Support Considerations 56
    - Pillar Design 56
    - Layout Modification For Horizontal Stress 57
    - Monitoring and Verification 57
- Summary and Conclusions 58
- References 59

Figures

Figure 1. Approximate location of underground stone mines for which pillar and roof span dimensions, rock mass characteristics, and excavation stability conditions were recorded. 2

Figure 2. Distribution of rock mass rating (RMR) values in stone mines obtained by direct classification of rock exposure in underground stone mines and laboratory testing of rock cores. 5

Figure 3. Partially benched pillar failing under elevated stresses at the edge of bench mining. Typical hourglass formation indicates overloaded pillar. The width-to-height ratio is 0.44 based on full benching height and average pillar stress of about 12% of the UCS. 11

Figure 4. Partially benched pillar that failed along two angular discontinuities. Width-to-height ratio is 0.58 based on full benching height; average pillar stress is about 4% of the UCS. 12

Figure 5. Pillar that had an original width-to-height ratio of 1.7, but failed by progressive spalling. Thin, weak beds are thought to have contributed to the failure. The average pillar stress was about 11% of the UCS prior to failure. 12

Figure 6. Remaining stump of a collapsed pillar in an abandoned area. Thin, weak beds in the pillar and moist conditions are thought to have contributed to the failure. The width-to-height ratio was 0.82 and average pillar stress about 11% of the UCS. 13

Figure 7. Example of rib spalling and resulting concave pillar ribs that can initiate when average pillar stress exceeds about 11% of the UCS. 14

Figure 8. Stable pillars in a limestone mine at a depth of cover of 275 m (900 ft). Slightly concave pillar ribs formed as a result of minor spalling of the hard, brittle rock. 14

Figure 9. Pillar that has been clad with chain link mesh to prevent further deterioration of the ribs. 15

Figure 10. Chart showing pillar performance based on a survey of 34 underground stone mines. 16

Figure 11. Effect of width-to-height ratio and rock mass rating (RMR) on pillar strength, based on numerical model results. 18

Figure 12. Sections through the center of pillars with different width-to-height ratios showing the extent of brittle and shear failure of the rock mass predicted by numerical modeling. 19

Figure 13. Example of a pillar that is bisected by a large, angular discontinuity. 20

Figure 14. Rib failure related to large, angular discontinuities adjacent to a fault zone. 21

Figure 15. Loss of pillar rib at the location of a large roof-to-floor discontinuity in a limestone mine. 21

Figure 16. Chart showing the impact of large, angular discontinuities on the strength of pillars, based on the results of numerical models. 23

Figure 17. Pillar damage observed in rock containing thin, weak bands. Note spalling of the intact rock material between the weak bands. 24

Figure 18. Stages of failure development in a beam of strong rock encased between two weak bands. 25

Figure 19. Example of bench mining of the floor between pillars in a limestone mine. 27

Figure 20. Stages of bench mining around a pillar used in the numerical models. 28

Figure 21. Results of numerical modeling showing strength reduction of pillars with initial width-to-height ratios of 1.0 and 1.5 from initial development through various stages of bench mining. Final width-to-height ratios at Stage 4 are 0.5 and 0.75. 29

Figure 22. Results of numerical model showing the average pillar stress during bench mining as a ratio of the average pillar stress prior to bench mining. 30

Figure 23. Change in the average vertical pillar stress and pillar strength during various stages of bench mining, for a pillar with a width-to-height ratio of 1.5 based on the results of numerical models. 31

Figure 24. Strength increase caused by increasing pillar length for pillars with various width-to-height ratios. Results from calibrated numerical models assuming rock failure initiates as spalling followed by shearing. 33

Figure 25. Chart showing the factor of safety against width-to-height ratio using equation 7. Current and disused pillar layouts are shown as well as single failed pillars. The recommended area for pillar design is shaded. 38

Figure 26. Distribution of roof span dimensions measured at 34 different underground stone mines. 41

Figure 27. Naturally stable 13.4 m (44 ft) wide roof span in a stone mine. 42

Figure 28. Bolts, straps, and injection grouting used to rehabilitate the roof at the site of a major roof fall. 43

Figure 29. Stability chart showing stone mine case histories and stability zones, modified and Diederichs [1996]. 46

Figure 30. Chart showing the effect of the thickness of the roof beam on excavation stability. ...47

Figure 31. Roof guttering at the pillar-roof contact. 48

Figure 32. Large stress-related, oval-shaped fall that has propagated upwards into weaker, overlying strata in a limestone mine. 49

Figure 33. Horizontal, stress-induced roof failure that initiated between two pillars. Arrows show measured direction of maximum horizontal stress. 49

Figure 34. Plan view showing the development of a stress-related roof fall in the direction perpendicular to the direction of the major horizontal stress, Iannacchione et al. [2003]. 50

Figure 35. Extension fractures exposed in a pillar rib after the pillar was bisected by a new crosscut. 51

Figure 36. Vertical cross section through a heading showing rock failure index values (a) without bedding discontinuities and (b) with a bedding discontinuity 1 m (3.3 ft) above the roofline. 52

Figure 37. Vertical cross section through a heading showing rock failure index values (a) with three 1-m thick (3.3 ft) bedding discontinuities in the roof and (b) thinly laminated roof. 52

Figure 38. Diagram showing room-and-pillar layout modified to minimize the potential impact of horizontal, stress-related damage. 53

Tables

Table 1. Uniaxial compressive strength of stone mine rocks collected at mine sites................... 3

Table 2. Summary of mining dimensions and cover depth of mines included in the study 9

Table 3. Characteristics of failed pillars. 10

Table 4. Summary of observed pillar instability associated with bench mining. 27

Table 5. Discontinuity dip factor (DDF) representing the strength reduction caused by a single discontinuity intersecting a pillar at or near its center, used in equation 5 35

Table 6. Frequency factor (FF) used in equation 5 to account for large discontinuities. 35

Table 7. Values of the length benefit ratio (LBR) for rectangular pillars with various width-to-height ratios..37

Acronyms and Abbreviations

DDF	discontinuity dip factor
FF	frequency factor
FOS	factor of safety
LBR	length benefit ratio
LDF	large discontinuity factor
NIOSH	National Institute for Occupational Safety and Health
RFRI	roof fall risk index
RMR	rock mass rating
UCS	uniaxial compressive strength
W:H	width-to-height ratio

Unit of Measure Abbreviations

ft	foot
in	inch
m	meter
mm	millimeter
MPa	megapascal
psi	pound-force per square inch

Pillar and Roof Span Design Guidelines for Underground Stone Mines

Gabriel S. Esterhuizen, Dennis R. Dolinar, John L. Ellenberger, and Leonard J. Prosser

Office of Mine Safety and Health Research
National Institute for Occupational Safety and Health

Background

Underground stone mines in the United States use the room-and-pillar method of mining in generally flat-lying bedded formations. Pillar and roof span stability are two essential prerequisites for safe working conditions in room-and-pillar mines. Unstable pillars can result in rock sloughing from the pillar and can lead to the collapse of the roof if one or more pillars should fail. In addition, the roof span between pillars must be stable to provide safe access to the working areas. Falls of ground from the roof and pillar ribs account for about 15% of all lost working days in underground stone mines [MSHA 2009]. In the past, pillar and roof span dimensions were largely based on experience at nearby mines, developed through trial and error or designed on a case-by-case basis by rock engineering specialists. This document presents the results of research conducted by the National Institute for Occupational Safety and Health (NIOSH) in cooperation with participating underground stone mines to develop generally applicable guidelines for designing stable pillars and roof spans in stone mines.

Research Methodology

An empirical approach was followed in developing these design guidelines. The actual performance of pillars and roof spans in 34 different stone mines in the Eastern and Midwestern United States was recorded. At each mine, the excavation dimensions, rock jointing characteristics, rock mass classification, and excavation stability conditions were recorded. In addition, rock samples were collected for strength testing at the NIOSH laboratory in Pittsburgh. Borehole scoping was conducted to observe the rock formation above the roofline at 13 mines. The collected data formed the basis for developing the design guidelines.

The pillar design guidelines were developed by selecting an appropriate strength equation and verifying it against the observed pillar performance. Modifications were made to the equation to account for geological structures and for the increased strength of rectangular pillars when compared to square pillars. Numerical models were used to assist in quantifying the strength modifications. The final safety factor and other design recommendations are based on the calculated safety factors of the observed pillar systems.

The roof span design guidelines are based on a pragmatic assessment of the current mining practices, roof span dimensions, and potential causes of roof instability. The survey of conditions in operating mines helped to identify the critical aspects of roof span stability. Stability analyses were conducted using empirical methods and numerical models to better understand the causes of instability. These design guidelines provide step-by-step procedure for minimizing the impact of potential sources of instability.

Applicability of Design Guidelines

The guidelines for pillar and roof span design are empirically based; their validity, therefore, is restricted to rock conditions, mining dimensions, and pillar stresses that are similar to those included in this study. These guidelines should be applicable to the majority of stone mines in the Eastern and Midwestern United States. If pillars need to be designed that are outside the validity of these design guidelines, the advice of rock engineering specialists should be sought.

Geotechnical Characteristics

Geological Setting

The stone mines included in this study are concentrated in the Interior Plains and the Appalachian Highlands physiographic regions [U.S. Geological Survey 2009]. Twenty-four of the mines fall within the Interior Plains region and are located in Illinois, Indiana, Iowa, Kentucky, Missouri, and Tennessee. The Appalachian Highlands region includes the remaining ten mines which are located in Maryland, Ohio, Pennsylvania, Tennessee, Virginia, and West Virginia. Figure 1 shows the approximate location of the mines included in the study.

Figure 1. Approximate location of underground stone mines for which pillar and roof span dimensions, rock mass characteristics, and excavation stability conditions were recorded.

Stone deposits located in the Interior Plains region are generally flat-lying or only gently dipping and include rocks ranging across most of the Paleozoic Era Ordovician Age to the Pennsylvanian Age. The Ordovician Age in this region includes the economic horizons of the Camp Nelson and the Tyrone limestones. The Mississippian Age includes the very siliceous and cross-bedded Loyalhanna as well as the Greenbrier and Monteagle limestones. The Monteagle is a gently dipping Upper Mississippian limestone which is mined on more than one horizon [Brann and Freas 2003]. In the Pennsylvanian Age, the Vanport member of the Allegheny Group is mined [Iannacchione and Coyle 2002].

Overall, the rocks encountered in the Appalachian Highlands region are similar in age to those found in the Interior Plains region. They differ from rocks in the Interior Plains region because they have been transformed through mountain building processes to consist of elongated belts of folded and faulted sedimentary rocks. Mines in the Appalachian Highlands region that were visited during this study operate in the Middle Ordovician Five Oaks formation, the Vanport and Loyalhanna formations mentioned previously, and the steeper dipping Monteagle formation which is mined at dips of up to 35°. Mines located in strata dipping greater than 10° were excluded from the study because of the complex loading conditions associated with increasing dip.

Intact Rock Strength

At each mine rock samples were collected to determine the uniaxial compressive strength (UCS). Cores were drilled from the samples and were tested at the NIOSH laboratory in Pittsburgh. The UCS results were grouped into three categories based on the average strength of rock samples obtained at the individual mine sites (shown in Table 1). The data show that there is considerable variation in the intact rock strength of the stone formations being mined at the various mines in this study.

Table 1. Uniaxial compressive strength of stone mine rocks collected at mine sites.

Group	Average MPa (psi)	Range MPa (psi)	Samples tested	Representative limestone formation names
Lower strength	88 (12,800)	44–143 (6,400–20,800)	50	Burlington, Salem, Galena-Plattesville
Medium strength	135 (19,600)	82–207 (11,900–30,000)	100	Camp Nelson, Monteagle, Plattin, Vanport, Upper Newman, Chickamauga
High strength	219 (31,800)	152–301 (22,000–43,700)	32	Loyalhanna, Tyrone

Rock Mass Characteristics

Discontinuities within the stone formations were subdivided into two groups: (a) near horizontal bedding-related structures and (b) steeply dipping joints or faults. All the sites visited contained at least one of the steeply dipping sets of joints. The average spacing of the steeply dipping discontinuities is 0.4 m (15 in) and the trace length was typically in the range of 1 to 3 m (3.3 to 10 ft). Joints are typically rough with no infilling or weathering. Isolated cases containing soft calcitic or clayey infill were observed. Large discontinuities that extend from the roof to the floor or across the width of an excavation were observed in about 40% of the locations visited. These are discussed in greater detail below.

Bedding layers do not always form a discontinuity in the rock. Many of the beds display a change in color without any substantial break in the continuity of the rock. Where bedding discontinuities exist, it was found that the trace length was greater than that of the steeply joint sets. Bedding discontinuities typically had very rough surfaces. Isolated cases of bedding joints with calcite or clayey infill were observed. The average spacing of bedding discontinuities is 0.9 m (3 ft) with trace lengths typically in the 3 to 10 m (10 to 30 ft) range with about 30% of the cases extending greater than 30 m (100 ft). Bedding discontinuities are often used to establish a stable roofline. It was found that 36% of the underground locations visited made use of a local bedding plane as the roofline.

Occasionally, bedding discontinuities were observed within the pillar ribs that extend over several hundred meters with relatively thick, weak clayey or calcite infill. Such bedding discontinuities are expected to have a significant effect on roof stability if they occur within the immediate roof of an excavation. Because such discontinuities are not visible when they are above the roofline, data on their presence are limited.

Large Joints

It was found that large, widely-spaced joints exist at about 40% of the underground sites visited. The average spacing of the large joints was 12 m (40 ft) with a minimum of 1 m (3.3 ft) and maximum of about 100 m (330 ft). The data collection approach used in this study did not identify spacings of larger than 100 m (330 ft). The dip of these discontinuities typically fell in the 70°–90° range, with isolated cases in the 30°–70° range. Large discontinuities that were parallel to the bedding were categorized as bedding-related features. The large discontinuities may contain soft infill materials, but the fill material is seldom more than 5 mm (0.2 in) in thickness.

Rock Mass Rating

In all cases the data collection for rock mass rating was conducted approximately 2 m (6 ft) from the floor of the mining horizon. The rating results, therefore, do not describe the detail of the rock layering in the immediate roof, but rather represent the typical rock mass conditions within the formation being mined. The rock mass ratings are presented in terms of the RMR-system [Bieniawski 1989] which classifies the rock mass on a scale of 0 to 100, with higher numbers indicating stronger rock masses. The RMR values fell within a narrow range, and the ratings for the immediate roof were not expected to be significantly different from the remainder of the formation. Figure 2 shows the distribution of RMR values obtained in this study. The values range from 60 to 85, which lay within the "good" to "very good" quality categories, according to the RMR classification tables [Bieniawski 1989].

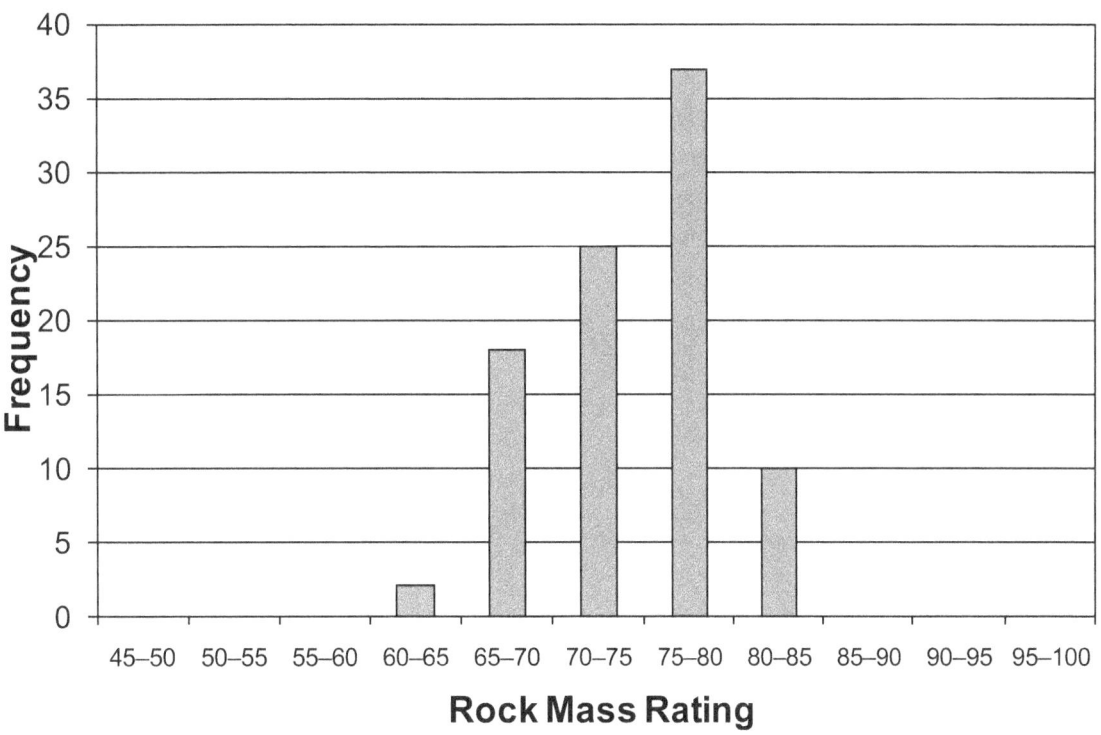

Figure 2. Distribution of rock mass rating (RMR) values in stone mines obtained by direct classification of rock exposure in underground stone mines and laboratory testing of rock cores.

Horizontal Rock Stress

Published stress measurements and field observations show that the horizontal stress in the Appalachian Highlands and Interior Plains regions can be much higher than the overburden stress. Horizontal stresses have been measured in various limestone mines [Iannacchione et al. 2003] and also in many of the coal mines in these regions [Mark and Mucho 1994]. Research has shown that horizontal stress may be explained by the effect of plate tectonics [Dolinar 2003; Iannacchione et al. 2002]. Tectonic loading is related to the movement of the North American Plate as it is pushed away from the Mid-Atlantic Ridge. A constant strain field from 0.45 to 0.90 millistrain is associated with tectonic loading, which induces higher horizontal stresses in the stiff limestone strata. The induced stress magnitude is not necessarily related to the cover depth for depths encountered in stone mining operations, but rather to the stiffness of the strata. Horizontal stresses are not necessarily present in all the stone formations because local features such as outcropping and folding may have relieved the stresses over geologic time [Iannacchione et al. 2003; Iannacchione and Coyle 2002].

A review of horizontal stress measurements in limestone and dolomite formations in the Eastern and Midwestern United States and Eastern Canada [Dolinar 2003] has shown that the maximum horizontal stress can vary from 7.6 MPa (1,100 psi) through 26 MPa (3,800 psi) up to depths of 300 m (1,000 ft). Limited information is available at greater depths. The orientation of the maximum horizontal stress for 80% of the sites in these regions is from N60°E to N90°E. This agrees with the regional tectonic stress orientation as indicated by the World Stress Map Project [2009]. The minimum horizontal stress is approximately equal to the vertical stress.

Pillar Design Considerations

Background

In a room-and-pillar mine, the pillars are required to provide global stability by supporting the overlying strata up to the ground surface. In addition, local stability in the form of stable ribs and roof spans between the pillars is required to provide safe working conditions. Pillar design is typically conducted by estimating the pillar strength and the pillar stress, and then sizing the pillars so that an adequate margin exists between the expected pillar strength and stress.

Pillar Strength

Pillar strength can be defined as the maximum resistance of a pillar to axial compression [Brady and Brown 1985]. In flat-lying deposits, pillar compression is caused by the weight of the overlying rock mass. Empirical evidence suggests that pillar strength is related to both its volume and its shape [Salamon and Munro 1967; Brady and Brown 1985]. Numerous equations have been developed that can be used to estimate the strength of pillars in coal and hard rock metal mines, and have been reviewed and summarized in the literature [Mark 1999; Martin and Maybee 2000; Lunder 1994; Hustrulid 1976]. Owing to the complexity of pillar mechanics, empirically based pillar strength equations, which are based on the observation of failed and stable pillar systems, have found wide acceptance [Mark 1999]. The empirical equations are only applicable for conditions similar to those under which they were developed. More recently, numerical model analyses combined with laboratory testing and field monitoring have contributed to the understanding of failure mechanisms and pillar strength [Lunder 1994; Iannacchione 1999; Mark 1999; Gale 1999; Kaiser et al. 2000; Diederichs et al. 2002; Esterhuizen 2006].

Pillar Stress Calculation

The average pillar stress (σ_p) in regular layouts of pillars can be estimated by assuming the overburden weight is equally distributed among all the pillars, known as the tributary area method:

$$\sigma_p = \gamma \times h \times \frac{(C_1 \times C_2)}{(w \times l)} \tag{1}$$

where γ = specific weight of the overlying rocks
h = the depth of cover
w = the pillar width
l = the pillar length
C_1 = the heading distance
C_2 = the crosscut center distance

This calculation generally provides an upper limit of the average pillar stress and does not consider the presence of barrier pillars or solid abutments that can reduce the average pillar stress. In conditions where the tributary area method is not valid, such as irregular pillars, limited extent of mining, or variable depth of cover, numerical models such as LaModel [Heasley and Agioutantis 2001] can be used to estimate the average pillar stress.

Pillar Failure

Pillar failure occurs when a pillar is compressed beyond its peak resistance and load shedding or yielding occurs [Brady and Brown 1985]. Failure of a single pillar can result in hazardous rib conditions, roof instability in the adjacent mining rooms, and blockage of local access ways. Load redistribution caused by the failure of a single pillar can overload the adjacent pillars, which can propagate into a wide-area failure [Salamon 1970; Zipf and Mark 1997]. These wide-area failures can occur as a catastrophic collapse within a few seconds or minutes or as a gradual "squeeze" over a number of hours or days. Wide-area collapses can cause excessive convergence of the mine opening, surface subsidence, or an air blast if they occur over a short period of time. Empirical evidence and theoretical studies suggest that as the width-to-height ratio of pillars is reduced, the potential for catastrophic failure increases as a result of the rapid decrease in strength of a slender pillar after it has reached its peak resistance [Salamon 1970].

Pillars can show signs of instability prior to failure. As the stress in a pillar increases, rock fracturing and spalling can occur at the pillar corners and can extend to the entire rib. Pillars that are stressed to the point of failure can exhibit an "hourglass" shape and ultimately develop open fractures and rib sloughing as the peak resistance is exceeded [Lane et al. 1999; Krauland and Soder 1987; Lunder 1994; Pritchard and Hedley 1993]. These signs of rock failure can be used to visually assess the stability of pillars in underground workings.

Safety Factor

The ratio of average pillar strength (S) to average pillar stress (σ_p) can be expressed as a factor of safety (FOS):

$$FOS = \frac{S}{\sigma_p} \qquad (2)$$

When designing a system of pillars, the FOS must be selected with care, because it must compensate for the uncertainty and variability inherent in the rock properties and mining dimensions. The selection of an appropriate FOS can be based on engineering judgment or statistical analysis of cases of both stable and failed pillars [Salamon and Munro 1967; Mark 1999; Salamon et al. 2006]. As the FOS decreases, the probability of failure of the pillars can be expected to increase. For example, a statistical analysis showed that a failure probability of 1:1,000 is associated with a FOS of 1.63 for coal pillars in Australian coal mines [Galvin et al. 1999], and a FOS of 1.0 is associated with 1:2 failure probability. In practical terms, if a few pillars are observed to be failed in a layout, it is an indication that the pillar stress is approaching the pillar strength, causing the weakest pillars in the layout to fail. The relationship between FOS and failure probability, however, depends on the uncertainty and variability of the system under consideration [Harr 1987].

Developing a Pillar Strength Equation for Stone Mines

The development of a pillar strength equation for stone mines followed a similar path as described above. The actual performance of pillars was observed in 34 different mines scattered throughout the Eastern and Midwestern United States. The observations of failed and stable pillars were used to identify the factors that were important to pillar strength. Numerical models were used to investigate some of the stability issues, such as the effect of large, angular discontinuities and the impact of weak bands within a pillar. The final strength equation and FOS recommendations are based on the analysis of the observed pillar performance in stone mines.

Survey of Stone Mine Pillar Performance

During this study, a survey of pillar performance in operating stone mines was conducted in the Eastern and Midwestern United States, where the majority of underground stone mining is conducted. Figure 1 shows the approximate location of the mines included in the survey. The measured pillar dimensions and depth of cover are summarized in Table 2.

Table 2. Summary of mining dimensions and cover depth of mines included in the study.

Dimension	Average	Minimum	Maximum
Pillar width	13.1 m (43.0 ft)	4.6 m (15.0 ft)	21.5 m (70.5 ft)
Pillar height	11.1 m (36.5 ft)	4.8 m (15.8 ft)	38.0 m (124.6 ft)
Width-to-height ratio	1.41	0.29	3.52
Room width	13.5 m (44.3 ft)	9.1 m (29.9 ft)	16.8 m (55.1 ft)
Intersection diagonal	21.7 m (71.2 ft)	16.1 m (52.8 ft)	29.6 m (97.1 ft)
Cover depth (ft)	117 m (385 ft)	22.8 m (75 ft)	670 m (2,200 ft)

Successful Pillar Layouts

The survey revealed that all 91 pillar layouts observed at the 34 different mines could be classified as successful in providing global stability, which is defined as supporting the weight of the overburden up to the ground surface. However, not all the pillar layouts were fully successful in providing local stability, which is defined as providing stable roof conditions and pillar ribs. The lack of local stability is generally managed by scaling of the roof and ribs or by installing appropriate support, and is not considered to be a failure of the pillar system.

Failed Pillars

A total of 18 cases of individual pillars that had failed in otherwise stable layouts were observed at five different mining operations. These failed pillars can represent a significant safety hazard because they are associated with unstable roof and ribs and typically require that the mining area be barricaded or abandoned.

Each of the failed pillars was visually assessed and, where possible, photographed to provide a record of the pillar conditions. The key parameters describing the failed pillars are summarized in Table 3, and include the probable factors contributing to the failure.

Table 3. Characteristics of failed pillars.

Case	Pillar width m (ft)	Pillar height m (ft)	Width-to-height ratio	Average pillar stress MPa (psi)	UCS MPa (psi)	Factors contributing to pillar failure
1	10.7 (35)	18.3 (60)	0.58	9.0 (1,305)	215.0 (31,175)	Partially benched pillar containing angular discontinuities
2	10.7 (35)	18.3 (60)	0.58	9.4 (1,363)	215.0 (31,175)	Partially benched pillar containing angular discontinuities
3	10.7 (35)	18.3 (60)	0.58	10.3 (1,494)	215.0 (31,175)	Partially benched pillar containing angular discontinuities
4	15.2 (50)	27.4 (90)	0.56	12.6 (1,827)	153.0 (22,185)	Pillar fully benched to 90 ft height causing reduced width-to-height ratio
5	10.7 (35)	18.3 (60)	0.58	12.8 (1,856)	215.0 (31,175)	Benched pillar, containing angular discontinuities
6	12.2 (40)	27.4 (90)	0.44	17.2 (2,494)	150.0 (21,750)	Partially benched pillar
7	8.5 (28)	15.9 (52)	0.54	17.2 (2,494)	150.0 (21,750)	Large, steep dipping discontinuity and elevated stress ahead of benching
8	12.2 (40)	27.4 (90)	0.44	17.3 (2,509)	150.0 (21,750)	Partially benched pillar
9	7.9 (26)	9.8 (32)	0.81	19.0 (2,755)	160.0 (23,200)	Thin, weak beds in limestone; pillar undersized causing elevated stress
10	12.8 (42)	7.3 (24)	1.73	17.4 (2,525)	160.0 (23,200)	Thin, weak beds in pillar causing progressive spalling
11	12.5 (41)	15.2 (50)	0.82	17.8 (2,583)	160.0 (23,200)	Thin, weak beds in pillar, moist conditions, and pillar collapsed
12	6.1 (20)	12.2 (40)	0.49	19.0 (2,755)	160.0 (23,200)	Benched pillar is undersized causing elevated stresses
13	6.7 (22)	12.2 (40)	0.54	20.0 (2,900)	160.0 (23,200)	Benched pillar is undersized causing elevated stresses
14	3.7 (12)	8.5 (28)	0.43	24.1 (3,495)	215.0 (31,175)	Undersized pillar subject to elevated stress
15	8.2 (27)	9.1 (30)	0.90	25.0 (3,625)	160.0 (23,200)	Thin, weak beds in pillar causing progressive spalling
16	5.5 (18)	7.3 (24)	0.75	27.0 (3,915)	160.0 (23,200)	Undersized pillar subject to elevated stress
17	12.2 (40)	15.9 (52)	0.77	8.4 (1,220)	164.8 (23,900)	Partially benched pillar containing angular discontinuities
18	12.2 (40)	15.9 (52)	0.77	7.6 (1,100)	164.8 (23,900)	Partially benched pillar containing angular discontinuities

The observed modes of pillar instability in stone mines can be divided into two categories. The first category is crushing failure which involves spalling and crushing of the solid rock with limited shearing along discontinuities such as joints or bedding planes. This failure mode is progressive and has been described in the following stages: (1) slight spalling of pillar corners and walls; (2) severe spalling; (3) appearance of fractures in the central part of the pillar; (4) occurrence of rock falls from the pillar and emergence of an hourglass shape; and (5) disintegration of the pillar, or, alternatively, the formation of a well-developed hourglass with the central section of the pillar completely crushed, Krauland and Soder [1987]. Rib spalling and emergence of an hourglass shape is the most common manifestation of crushing failure in stone mines, as shown in Figure 3.

The second category of pillar instability is structure-controlled failure which is characterized by shearing along geologic discontinuities, such as large through-going joints or faults, or weak bedding planes. Pillars that are intersected by large through-going discontinuities, as shown in Figure 4, can fail if sliding occurs along the discontinuity. Weak bedding planes that contain soft infill materials can extrude and loosen the rock or induce fracturing of the adjacent intact rock, which can result in progressive disintegration of the pillar. The pillar shown in Figure 5 and the totally collapsed pillar shown in Figure 6 both appear to have failed in this manner.

The observed failed pillars were typically surrounded by pillars that appeared to be stable, showing minimal signs of disturbance. The observations lead to the conclusion that the failed pillars represent the low end of the distribution of possible pillar strengths, and not the average pillar strength.

Figure 3. Partially benched pillar failing under elevated stresses at the edge of bench mining. Typical hourglass formation indicates overloaded pillar. The width-to-height ratio is 0.44 based on full benching height and the average pillar stress is about 12% of the UCS.

Figure 4. Partially benched pillar that failed along two angular discontinuities. Width-to-height ratio is 0.58 based on full benching height; average pillar stress is about 4% of the UCS.

Figure 5. Pillar that had an original width-to-height ratio of 1.7, but failed by progressive spalling. Thin, weak beds are thought to have contributed to the failure. The average pillar stress was about 11% of the UCS prior to failure.

Figure 6. Remaining stump of a collapsed pillar in an abandoned area. Thin, weak beds in the pillar and moist conditions are thought to have contributed to the failure. The width-to-height ratio was 0.82 and average pillar stress about 11% of the UCS.

Observed Rib Instability

Rib spalling is one of the early signs of elevated pillar stress. Figure 7 shows an example of rib spalling at approximately 270 m (900 ft) of cover. Spalling is characterized by fractures through the intact rock that are parallel to the direction of the maximum stress. Spalling normally initiates at the pillar corners and can spread to the pillar ribs, resulting in slightly concave ribs, shown in Figure 8. Rib spalling was observed to initiate when the average pillar stress exceeds about 11%–12% of the UCS. However, not all pillars that exceeded the 11%–12% stress ratio showed signs of rib spalling. It should, therefore, be interpreted as the lower limit for the onset of rib spalling. Rib instability can additionally be caused by unfavorable jointing in the rock mass or by poor blasting practices. The hazard associated with rib spalling can be mitigated by barring-down the loosened material, but this has the detrimental effect of reducing the pillar size. In some cases mine operators installed rib support, such as chain link mesh and bolts, to secure the rib walls, as shown in Figure 9.

Figure 7. Example of rib spalling and resulting concave pillar ribs that can initiate when average pillar stress exceeds about 11% of the UCS.

Figure 8. Stable pillars in a limestone mine at a depth of cover of 275 m (900 ft). Slightly concave pillar ribs formed as a result of minor spalling of the hard, brittle rock.

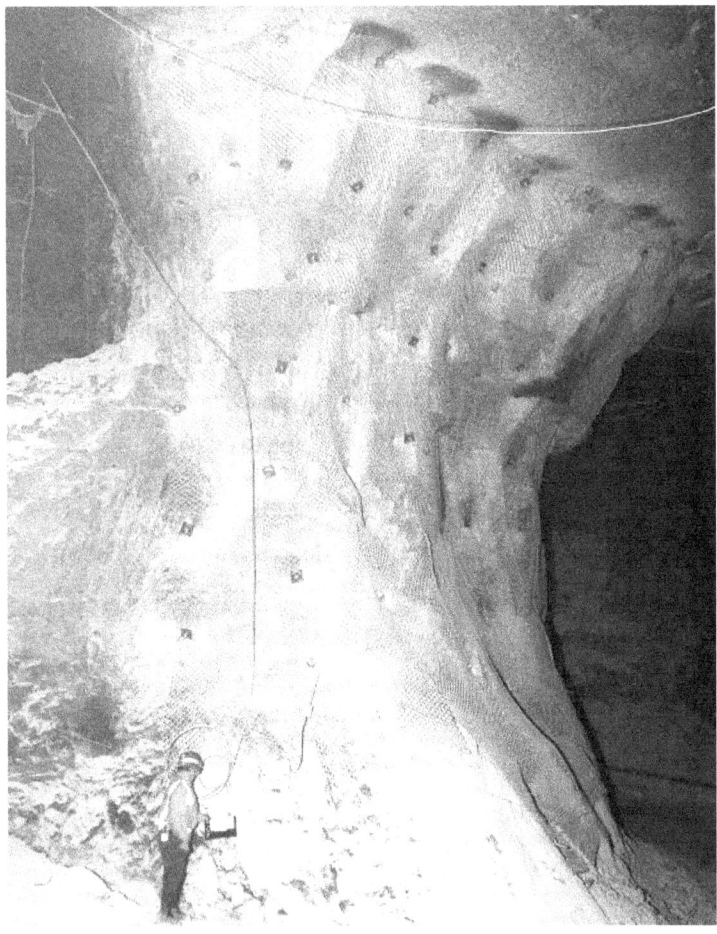

Figure 9. Pillar that has been clad with chain link mesh to prevent further deterioration of the ribs.

Wide-Area Failures

None of the mines that were included in the survey had experienced wide-area pillar failures, in which multiple pillars had failed. However, two cases of wide-area pillar failure were reported in limestone mines that are no longer operational [Zipf 2001]. The first case was a reported collapse of a small stone mining operation that may have been the result of a sudden collapse of the pillars. The pillar dimensions were variable and insufficient information exists to evaluate this event for estimating pillar strength.

The second case was a failure in which an area of about 20 pillars was reported to have failed [Zipf 2001]. An investigation of this failure revealed that the pillars had not failed, but moisture-related yield of the weak floor may have occurred that triggered the surrounding roof to collapse around the pillars [Zipf 2008]. The pillars were seen to be intact within the collapsed area. Consequently, this case has been discounted for estimating pillar strength because the pillars had not failed. However, it does highlight the fact that the potential for yielding floor should be evaluated when designing a stone mine pillar layout (e.g., by drilling into the floor during exploration).

These two case histories, while not directly useful for evaluating stone mine pillar strength, do emphasize the fact that wide-area pillar failures can and have occurred in U.S. stone mines.

Summary Chart of Pillar Observations

The pillar layouts that were surveyed by NIOSH are presented in Figure 10 which shows the normalized pillar stress against the width-to-height ratio. The pillar stress is normalized by the average UCS of the intact rock (obtained from Table 1). Figure 10 also includes data points representing the 18 failed pillars (that are presented in Table 3), the failures associated with the presence of large, angular discontinuities, information on the approximate number of pillars in each layout and indicates whether a pillar layout is no longer in use. A bounding curve was drawn around the case histories, which represents the limit of current experience with stone mine pillar performance.

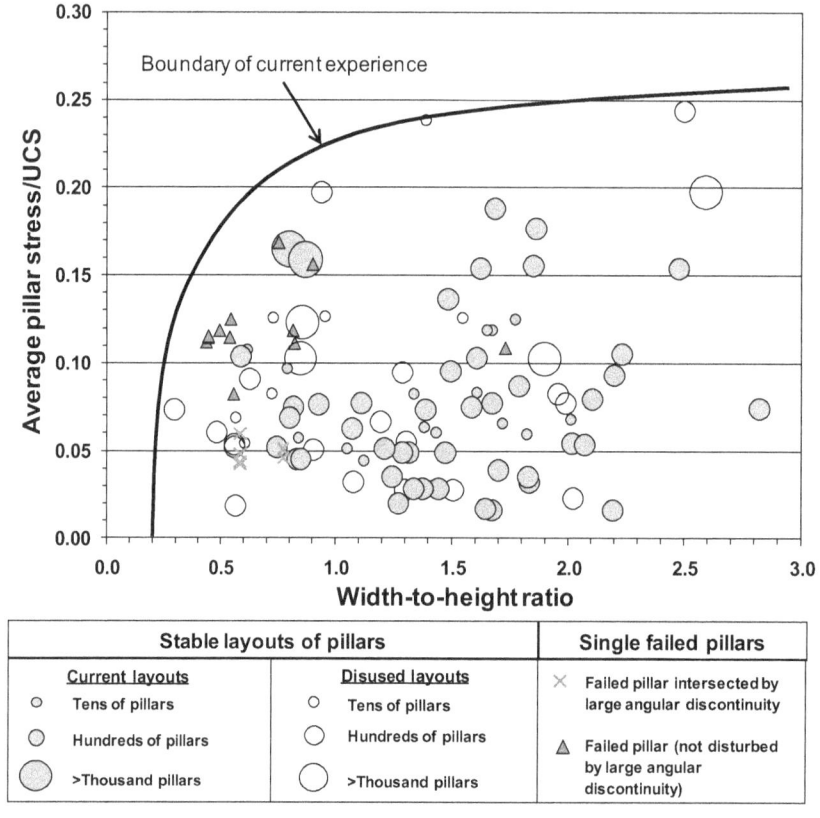

Figure 10. Chart showing pillar performance based on a survey of 34 underground stone mines.

For the purpose of preparing this chart, the width-to-height ratio of the pillars was based on the minimum pillar width. Where pillars were partially benched, the full height of benching was used to represent the pillar height. Actual underground measurements of room-and-pillar dimensions were used.

All the pillar layouts shown in the chart (i.e., both current layouts and those no longer in use) can be considered to have been "successful" in the primary objective of providing support to the overburden. The results show that these successful pillar layouts contain many thousands of stable pillars while the failed pillars are all single cases that represent only a very small part of the total population of pillars. The relatively low strength of the failed pillars that contained angular discontinuities is also clearly indicated. The chart can be used to compare a current or proposed pillar layout with past experience [Esterhuizen et al. 2008].

Stone Mine Pillar Stability Analysis

The survey of pillar performance in operating mines helped to identify the potential causes of pillar instability. However, the impact of variations among these factors could not be quantified by observations alone. Further analyses were conducted to investigate some of the identified stability issues and other aspects of pillar stability that need to be understood when designing pillars. The first issue evaluated below is the impact of brittle rock spalling on the slender pillars encountered in stone mines. This is followed by an evaluation of the impact of large, angular discontinuities and weak bedding bands on pillar strength. Finally, the effects of floor benching between pillars and the effects of increasing the length of a pillar on its strength are evaluated.

Brittle Rock Spalling

The hard rock that is extracted by stone mines (such as limestone, dolomite, and sandstone), can be classified as brittle rock, owing to the tendency of this type of rock to rapidly lose strength after the peak load-bearing capacity of the rock has been reached. Failure of the rock surrounding underground excavations in hard, brittle rock tends to initiate by a process of spalling in which slabs of rock are formed parallel to the excavation surfaces. Spalling failure was observed in several stone mines. Spalling is a process that occurs when the confining stress is low and the rock splits in a direction parallel to the major compressive stress and forms slabs which can dislodge and fall [Stacey, 1981]. Assessment of the spalling mode of failure [Martin and Chandler 1994; Diederichs et al. 2002] shows that extension fractures [Stacey 1981] develop at low confinement, which can be seen as a cohesion weakening process [Hajiabdolmajid et al. 2000]. As the confining stress increases, the frictional properties of the rock are mobilized resulting in resistance to shearing. Spalling can initiate at a stress that is much lower than the uniaxial compressive strength of the rock [Kaiser et al. 2000; Diederichs 2002; Stacey and Yathavan 2003]. For example, spalling in stone mines appears to start when the average pillar stress is only about 10% of the UCS of the rock.

The pillars used in stone mines tend to be relatively slender when compared to pillars used in most other mining applications. For example, the average width-to-height ratio of the pillars observed in stone mines was 1.41 with a minimum of 0.29. Slender pillars behave differently from wider pillars because of the absence of a confined core. In wide pillars, the central core of the pillar is confined by the perimeter material which results in an increase in the overall strength of the pillar. When pillars are slender, this confinement is absent or may be insignificant, resulting in lower pillar strength. Numerical analysis of generic hard rock pillars [Esterhuizen 2006] seems to indicate that there is little change in pillar strength when the width-to-height ratio drops below 1.0. However, experience has shown that very slender pillars can be expected to be weaker than predicted by the models because of the increasing importance of local discontinuities on pillar stability. As the width-to-height ratio increases beyond 1.0, the pillar strength will increase rapidly as confinement is generated [Lunder 1994]. Figure 11 shows numerical model results [Esterhuizen 2006] in which both the spalling and shearing failure modes were modeled. The results show how the width-to-height ratio and the rock mass rating affect pillar strength. In these models it was assumed that spalling occurred when the maximum stress was one third of the UCS, which is higher than observed in limestone formations, and the confining stress is less than 10% of the maximum stress. It can be seen that the pillar strength only starts to respond to the increasing width of the pillar when the width-to-height ratio is greater than 0.7. The results also show that the rate of increase in strength is related to the rock mass rating.

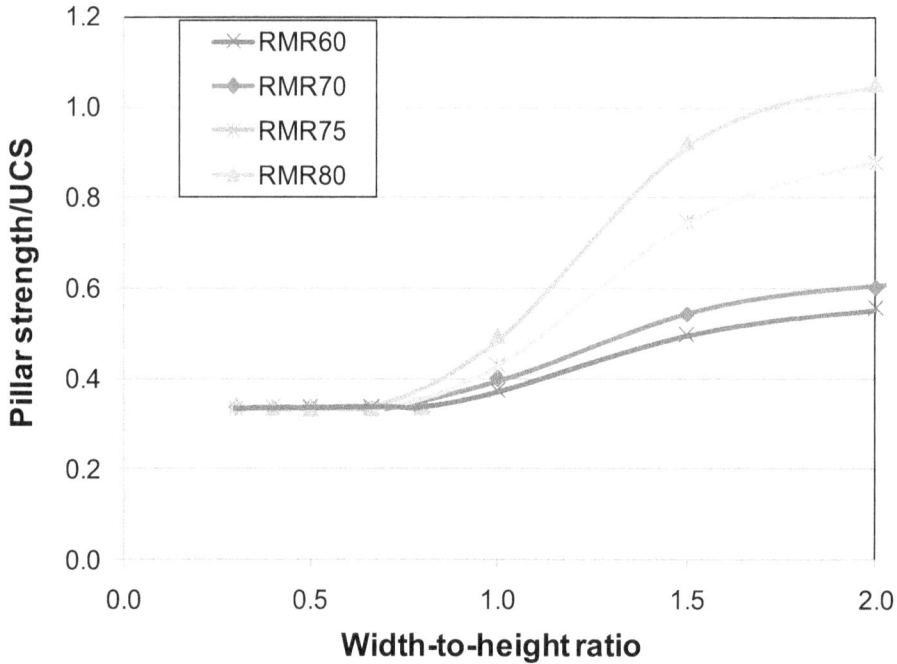

Figure 11. Effect of width-to-height ratio and rock mass rating (RMR) on pillar strength, based on numerical model results.

Another issue affecting the strength of a slender pillar is the potential for spalling failure to progress through the pillar. When pillars are wide, spalling failure typically starts at the perimeter of the pillar while the central core remains intact and provides resistance to the imposed stress. As the stress increases, a wide pillar will progressively fail from the outside inwards. The failed perimeter of the pillar provides confinement to the core, which allows the shear strength of the rock to develop. The ultimate pillar strength will depend on both the spalling and shearing strength of the rock. However, when pillars are slender, the spalling mode of failure can extend through the pillar and the higher shearing strength is not mobilized. Figure 12 shows numerical model results in which the spalling-shearing failure modes were simulated. When the width-to-height ratio is 0.5, the pillar fails entirely by spalling. When the width-to-height is increased to 1.0, some shearing failure occurs which produces a small increase in the pillar strength. When the width-to-height ratio is increased to 2.0, shearing becomes the dominant failure mode and there is a further increase in the pillar strength.

The stress-strain characteristics of the modeled pillars also showed that slender pillars start spalling when they are at or near their peak strength; conversely, wider pillars start spalling well before they reach their peak strength. This implies that if slender pillars show signs of spalling, they may be loaded at or near their ultimate strength, and failure may be imminent. The slender pillars also display a rapid drop in strength after reaching the peak strength, and the wider pillars lose their strength more gradually. A rapid drop in strength can result in violent pillar failure [Salamon 1970].

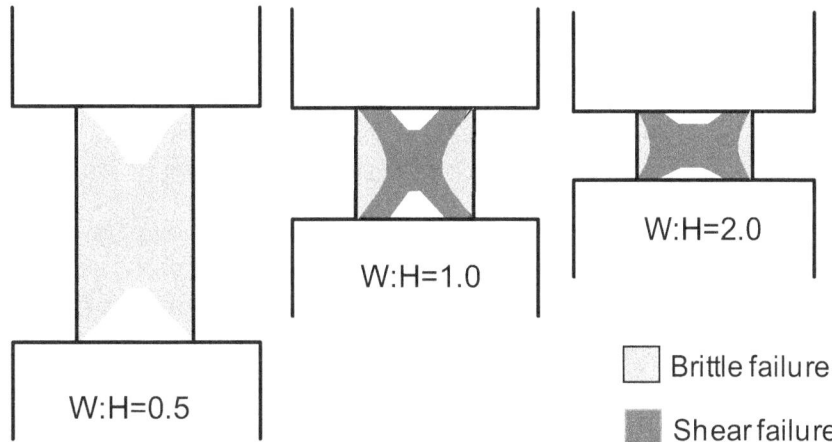

Figure 12. Sections through the center of pillars with different width-to-height ratios showing the extent of brittle and shear failure of the rock mass predicted by numerical modeling.

In summary, this study of the impact of brittle rock spalling illustrated that:

- The brittle failure process identified in stone mine pillars is common in hard rock mines and can occur at stress magnitudes that are well below the UCS of the rock.
- Observations indicate that brittle fracturing and spalling can start when the average pillar stress is only about 10% of the UCS of the rock.
- The lack of confinement in slender pillars that have width-to-height ratios of less than 1.0 can imply that brittle fracture will occur completely through these pillars at relatively low stress magnitudes.
- Slender pillars are more prone to sudden failure because they lose their strength rapidly once they are overloaded.
- These results indicate that it would be prudent to avoid using excessively slender pillars in stone mine design, especially if the stress magnitude is expected to result in brittle fracture and spalling of the intact rock.

The Impact of Large, Angular Discontinuities

Large, angular discontinuities were observed to have contributed to the failure of 7 of the 18 pillar failures listed in Table 3. Large discontinuities were observed to be present in 22 of the 34 stone mines surveyed. Pillars failures associated with angular discontinuities occurred when the average pillar stress was only about 5% of the UCS. The potential weakening effect of a large angular discontinuity is clearly demonstrated in Figure 13, which shows that sliding of the upper part of the pillar over the lower part can easily occur. These discontinuities are not always readily visible to production staff when developing a pillar, but only become apparent when the pillar becomes fully loaded or when bench mining is conducted around the pillars. Particularly hazardous conditions can result if large angular discontinuities cause unstable blocks of rock to slide or topple from the pillar ribs, as shown in Figures 14 and 15.

These large discontinuities can be widely spaced, extend from the roof to the floor of the workings, and the extent of the strike can be several hundred feet. The spacing appears to follow a negative exponential distribution with 75% of the discontinuities less than 12 m (40 ft) apart. The average dip was 81°; only 18% of the discontinuities observed in this study had a dip less than 70°.

Figure 13. Example of a pillar that is bisected by a large, angular discontinuity.

Figure 14. Rib failure related to large, angular discontinuities adjacent to a fault zone.

Figure 15. Loss of pillar rib at the location of a large roof-to-floor discontinuity in a limestone mine.

Studies of the impact of large roof-to-floor discontinuities on pillar strength have been conducted by using numerical models and then comparing the results to actual pillar performance [Esterhuizen 2000, 2006]. In these studies, a series of numerical models were created to simulate pillars with a variety of width-to-height ratios; each model contained a large discontinuity intersecting the center point of the pillar, with the strike of the discontinuity parallel to the pillar edges. The friction angle of the discontinuities was set at 30°, and each discontinuity was assumed to have no cohesive strength. The intact rock was modeled to simulate a typical limestone formation displaying brittle spalling at low confinement. Various analyses were conducted in which the dip of the discontinuity was varied from 30° to 90°, and the strength of the pillar was determined by gradually compressing the pillar until it failed. A series of curves were fitted to the model results and are shown in Figure 16. The results show that, as the discontinuity dip increases from 30° to about 60°, its impact on the pillar strength increases; but, when the discontinuity dip is greater than 70°, the impact on pillar strength starts to diminish. A vertical joint through the center of a pillar was shown to have a relatively small impact on pillar strength.

The width-to-height ratio is also shown to be a significant factor in the impact of large discontinuities. The graph in Figure 16 shows, for example, that a pillar with a width-to-height ratio of 0.5 will suffer a 95% reduction in strength if it is intersected by a 60° joint, and a pillar with a width-to-height ratio of 1.0 would only suffer a 34% reduction in strength. Smaller, angular discontinuities within the rock mass can be expected to have a similar but less severe impact on the strength of slender pillars. The sensitivity of slender pillars to the presence of angular discontinuities is further motivation for avoiding such slender pillars when designing a mine layout.

Observations of failed pillars confirm that large strength reductions can occur when large, angular discontinuities are present. The observation that slender pillars fail at about 5% of the intact rock strength can be explained by the impact of large discontinuities. If large discontinuities are likely to be found, increasing the width-to-height ratio of the pillars is probably the most effective method of achieving greater pillar strength.

The field observations and numerical model studies have shown that:
- Large, angular discontinuities can cause a significant reduction in the strength of pillars and should be accounted for in pillar design.
- The strength reduction caused by large, angular discontinuities is most severe in tall, slender pillars; the severity of the strength reduction decreases as the width-to-height ratio increases.
- Large, angular discontinuities were present in about 65% of the mines surveyed. Due to this finding, the presence or absence of these structures should be verified during geotechnical assessments.

These findings are a further confirmation that excessively slender pillars should be avoided. The pillar strength calculation procedure, described in the Pillar Strength Equation for Stone Mines section, takes into account the weakening effect of large, angular discontinuities.

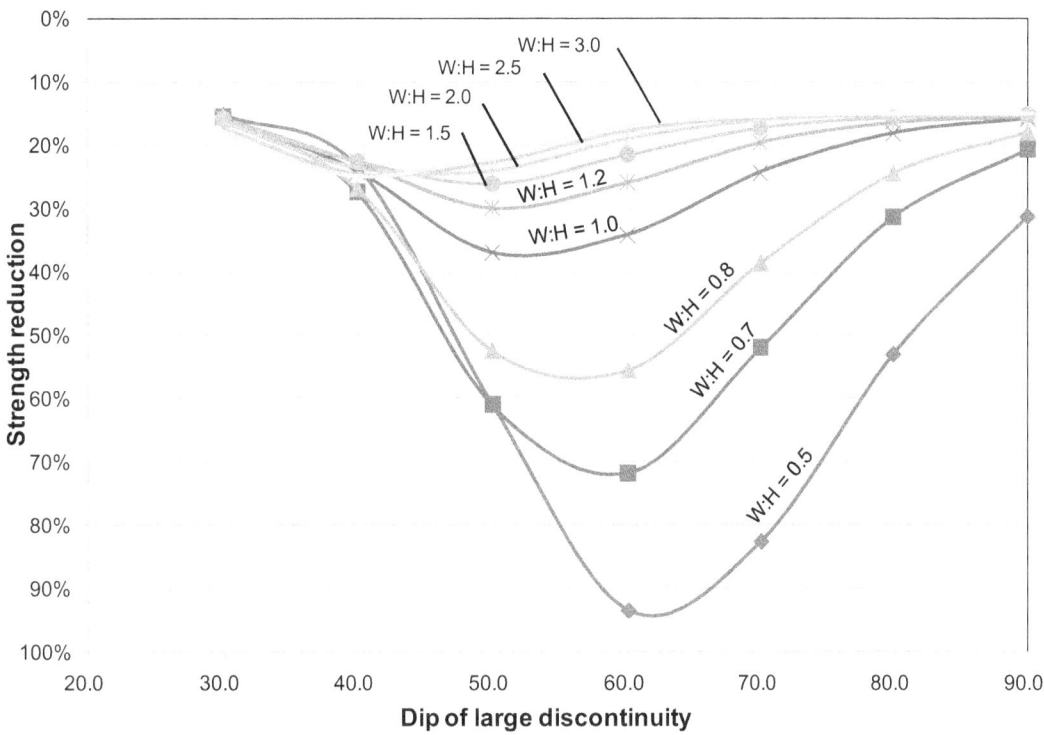

Figure 16. Chart showing the impact of large, angular discontinuities on the strength of pillars, based on the results of numerical models.

The Impact of Weak Bedding Bands

The presence of near-horizontal, thin, weak bands within a pillar was identified as a contributing factor in four of the failed pillar cases. The material comprising the weak bands observed in the field included carbonaceous bedding planes, calcite fillings on bedding planes, and indurated clays or seat earths that had characteristics more closely resembling soils than rocks.

Based on the field observations, it appeared that extrusion of the weak bands contributed to the failure of the stronger limestone. At lower loads, the soft material would extrude and release blocks of intact rock defined by preexisting joints. This typically caused overhangs in the pillar ribs. At higher vertical loads, the intact rock appeared to fracture into thin, vertical slabs. It was speculated that this failure was related to horizontal tensile stresses that develop as the weaker material extrudes under the elevated loads. Similar mechanisms of failure have been documented in other cases [Brady and Brown 1985; Hoek et al. 1995].

The pillars shown in Figures 5 and 6, listed as case 10 and 11 in Table 3, are examples of pillars that appear to have failed by this mechanism. In both cases the average pillar stress was only about 11% of the UCS of the limestone beds at this mine. Other pillars at the same mine seemed to be at an early stage of the same failure mechanism, where spalling is associated with the presence of weak, soft bands within the pillar. Figure 17 is an illustration of a partially failed pillar that appears to have failed by the same mechanism.

Figure 17. Pillar damage observed in rock containing thin, weak bands. Note spalling of the intact rock material between the weak bands.

A previous study investigated the mechanism of failure and the impact of weak bands on pillar strength [Esterhuizen and Ellenberger, 2007]. Numerical models were used to simulate the stress and associated rock failure in a slab of strong rock encased between two weaker bands. The strong rock was modeled as a hard, brittle material, with spalling behavior; while the weak bands were modeled as a low-cohesion, soil-like material. The model results showed, for example, that a uniform rock slab model, consisting of only the stronger rock material, had a strength of 35 MPa (5,000 psi), which is approximately equal to the brittle strength of the material. When weak bands are added, the strength can be as low as 6.8 MPa (990 psi).

Inspection of the model outputs showed that failure of the layered rock mass occurred through an extrusion-tension mechanism. As the vertical load is increased, failure first occurs in the weak bands because of their low strength. As the load continues to increase, a zone of tension develops within the stronger slab (Figure 18a), which is caused by the extrusion of the failed weak bed material. As the loading increases, the tensile stresses increase and tensile failure develops in the stronger slab (Figure 18b), which relieves the initial zone of tension (Figure 18c). As the vertical loading continues to increase, tensile stresses are induced on either side of the initial tensile failure zone and tensile failure continues to occur. The process repeats until the entire slab has failed or the extrusion mechanism is inhibited by frictional resistance between the weak bands and the rock slab. If the tensile failure process is inhibited, the remainder of the slab fails by shearing. The extrusion-tensile failure mechanism can explain the observed progressive spalling of intact rock at relatively low stress.

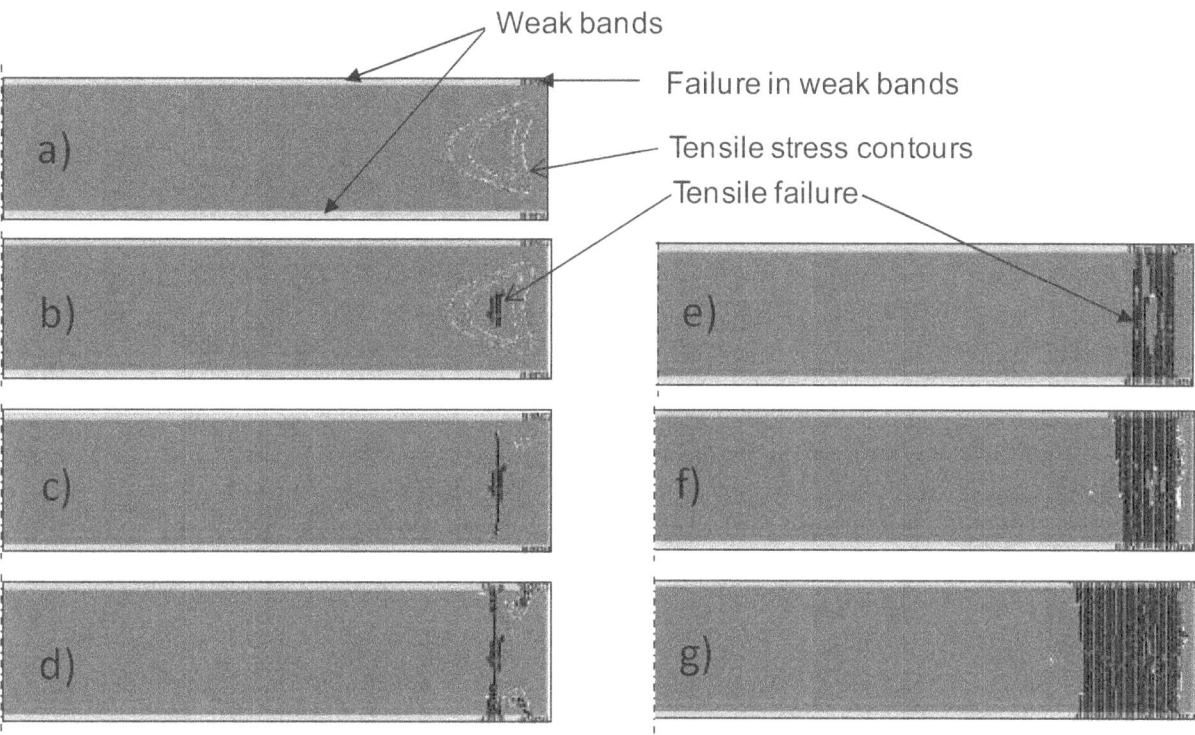

Figure 18. Stages of failure development in a beam of strong rock encased between two weak bands.

The sensitivity of the weak-bedded rock mass to variations in the elastic modulus of the weak bands, the strength of the weak bands and the tensile strength of the stronger material was tested. The results showed that the extrusion-tensile failure mode occurred for most scenarios. However, as the weak bands became stronger and stiffer, the role of tensile failure was diminished and shear failure of the strong rock slab became more prevalent. Conversely, when weak bands are thick and soft, extrusion can occur without inducing tensile failure in the stronger rock.

The results of the field observations and numerical model studies showed that:

1. The failure mechanism in a pillar with weak bands is predominantly caused by extrusion of the weak bands, which induces tension in the stronger rock slabs. The strong rock fails due to tension, which is manifested as rib spalling in underground stone mines.
2. The extrusion-tension failure mechanism can cause a significant reduction in the strength of the rock mass.
3. The extrusion-tension failure mode typically initiates at the perimeter of a pillar and progresses inwards, reducing the effective width of the pillar.
4. Observations in operating mines show that weak bands can cause rib failure to initiate when the average pillar stress is only about 10% of the rock strength.
5. At lower stresses the extrusion process can release blocks defined by joints or blasting fractures.
6. Slender pillars (width-to-height<1.0) are more severely affected by the presence of weak bands than wider pillars.

At present there is insufficient information to clearly identify the conditions that might lead to the extrusion-tension mode of failure or to predict the impact of weak beds on pillar strength in a generally applicable manner. It is not clear, for example, why only the single pillar shown in Figure 6 collapsed while the rest of the pillars in the area did not show signs of distress, although they appeared to have similar weak bands. For these reasons, pillars that contain weak beds were excluded from the pillar strength estimation method described in this document. In such cases, it should be noted that weak beds can cause a significant reduction in the strength of the pillars, and a detailed geotechnical investigation by a rock engineering specialist should be conducted.

The Effect of Floor Benching

Bench mining of the floor between pillars is a common practice in stone mines where the formation thickness exceeds the practical height of initial development mining. Bench drilling equipment from open pit mines is often used resulting in a highly efficient method of production. Figure 19 shows a floor bench with partially benched pillars on the upper level and fully benched pillars in the foreground. Observations showed that pillars can become unstable at the edge of the benching operations. Table 4 summarizes the cases in which pillar instability was associated with bench mining. Benching operations were halted in two of the observed cases owing to instability of the partially benched pillars. Several cases were observed in which the pillars at the perimeter of the benching area showed signs of increased loading. In addition, instability was observed when bench mining exposed large joint structures in the pillar ribs.

Figure 19. Example of bench mining of the floor between pillars in a limestone mine.

Table 4. Summary of observed pillar instability associated with bench mining.

Case	Development width-to-height ratio	Benched width-to-height ratio	Average pillar stress MPa (psi)	Instability observed
1	1.30	0.59	13.1 (1,900)	Large discontinuities exposed by benching; diagonal shearing through pillar. Benching was halted.
2	1.50	0.73	14.1 (2,045)	Progressive spalling of pillar ribs; pillar width reduced significantly; weak bedding infill contributed to spalling.
3	1.50	0.44	15.0 (2,175)	Spalling of several pillars caused hourglass shape. Sloughing from one of the pillars caused by a large, steeply dipping discontinuity.
4	1.65	0.61	8.1 (1,175)	Sloughing from pillar ribs after a large discontinuity is exposed at the perimeter of benching.
5	2.00	0.99	19.8 (2,871)	Sloughing from pillar walls at location of a large discontinuity. Benching was halted and resumed beyond this area.
6	1.96	0.92	13.1 (1,900)	Spalling caused hourglass shape. Benching halted owing to presence of large discontinuities in adjacent pillars.

Three-dimensional, numerical model analyses were conducted to further investigate the likely load and strength changes caused by benching and to evaluate their impact on pillar stability. Details of the analyses are presented in Esterhuizen et al. [2007]. The analyses were based on the assumption that rock failure initiates by a process of brittle spalling. The change in pillar strength and stress was evaluated as the pillar height was progressively increased by benching. Figure 20 shows various stages of benching around a pillar; these are the stages that were used in the models.

Analyses were conducted for pillars with initial width-to-height ratios of 1.0 and 1.5. The models simulated benching that doubled the height of the pillars. The model results presented in Figure 21 shows that the strength of the pillar with a width-to-height of 1.0 is reduced by about 16% from its initial value of 48 MPa (6,900 psi) to a final value of 40 MPa (5,800 psi). The pillar with a width-to-height ratio of 1.5 experiences a reduction in strength of 37% from the development stage to the fully benched stage. One reason for the smaller strength reduction in the narrow pillar with a width-to-height ratio equal to 1.0 is that the strength is already near the minimum value on development, and increasing the height during benching only causes a small additional reduction in strength.

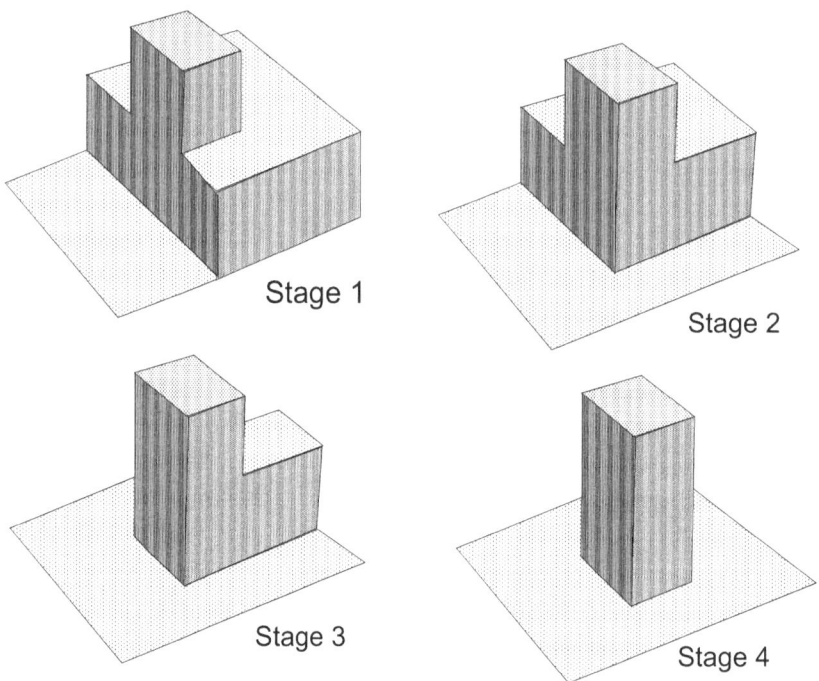

Figure 20. Stages of bench mining around a pillar used in the numerical models.

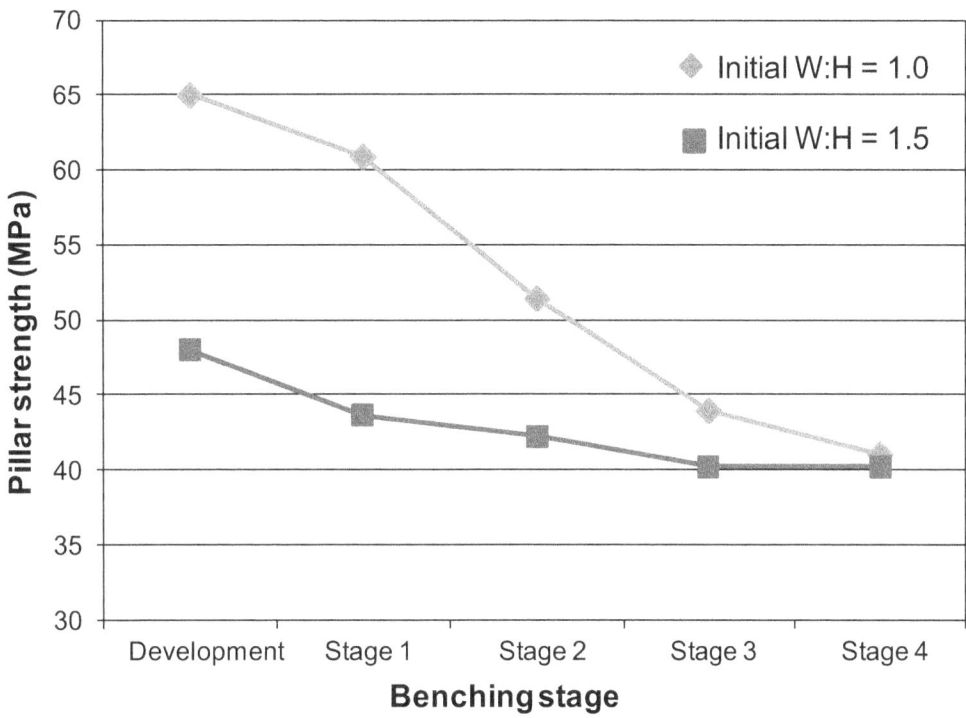

Figure 21. Results of numerical modeling showing strength reduction of pillars with initial width-to-height ratios of 1.0 and 1.5 from initial development through various stages of bench mining. Final width-to-height ratios at Stage 4 are 0.5 and 0.75.

From a pillar design point of view, it is important to also know how bench mining affects the pillar loads while the strength reduction is occurring. Figure 22 shows the average stress in the pillars near benching operations, as obtained from numerical models. The results show that the development pillars at the edge of the benched area are subject to an increase of about 12% in their average stress. The results further show that the stress in the partly benched pillar is lower than the stress in the adjacent pillar that has not been benched yet, indicated as "Perimeter pillar." The lower stress in the partially benched pillar can be explained by the fact that its stiffness is reduced by the increase in height of one of the sides of the pillar, causing the load to be transferred to the stiffer development pillars. It can clearly be seen that the fully benched pillars are at a reduced stress level, owing to their relatively low stiffness. As benching continues, the stress in the fully benched pillars is expected to gradually increase back to the tributary stress.

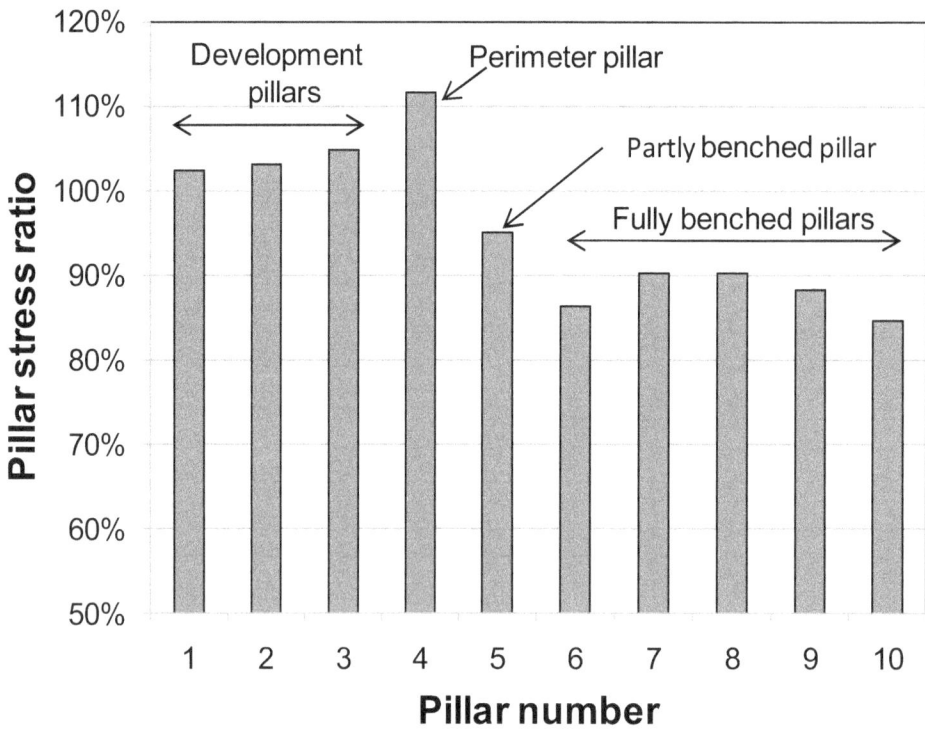

Figure 22. Results of numerical model showing the average pillar stress during bench mining as a ratio of the average pillar stress prior to bench mining.

The changes in both the average vertical stress and the pillar strength are presented in Figure 23 for a pillar with a width-to-height ratio of 1.5. The pillar stress is shown to reach a maximum value just before benching starts around the pillar. As soon as one side of the pillar has been benched, the average pillar stress decreases, owing to the increased height and reduced stiffness. The average pillar stress continues to decrease as benching progresses, until the pillar is fully benched. The stress in the fully benched pillars will gradually rise as the benching face moves away. Full tributary loading can reestablish in the benched pillars if the mined area is sufficiently large.

The numerical models confirm that elevated stresses can occur in pillars around the perimeter of a benched area. However, the existence of reduced stresses in the partially benched pillars seems to be in conflict with the field observations, which indicate that elevated stresses exist in the partially benched pillars. Closer inspection of the model results show that stresses are not symmetrically distributed within pillars at the edge of a benched area. Zones of high stress exist within the partially benched pillars; these are likely to contribute to the failure observed in the partially benched pillars in operating mines.

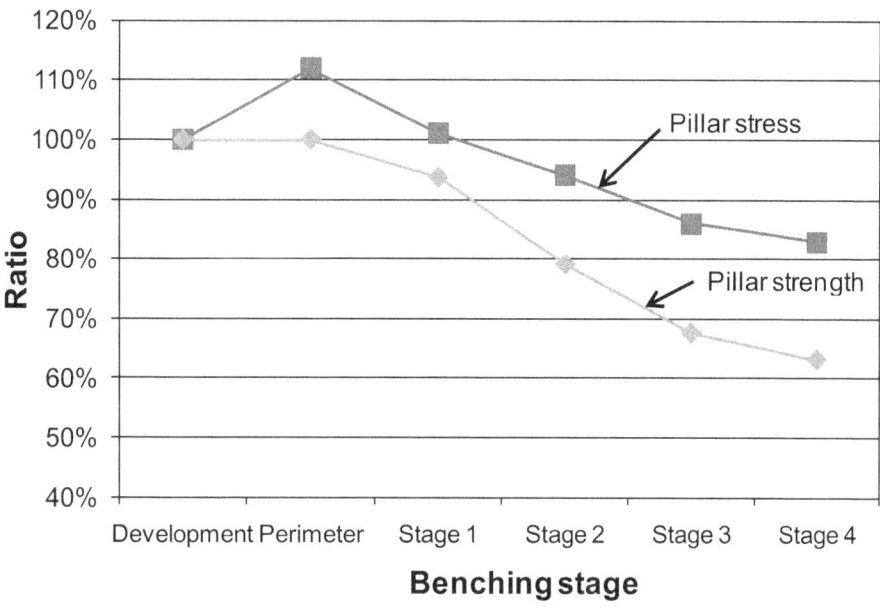

Figure 23. Change in the average vertical pillar stress and pillar strength relative to development during various stages of bench mining, for a pillar with a width-to-height ratio of 1.5 based on the results of numerical models.

These results of the field observations and numerical model analyses indicate that:

1. Instabilities in benched pillars can be caused either by geological structures in the rock or by an increase in pillar stress.
2. Large, angular joints or other geological discontinuities are more likely to be exposed by the increased height of benched pillars.
3. The numerical model results showed that the stress increase at the benching line is probably caused by the difference in stiffness of the benched and development pillars.
4. The numerical models showed that there was an increase in stress of about 15% in the pillars around the perimeters of the benching operations.
5. The numerical model results showed that the strength of a pillar is reduced in a near linear manner as each side of the pillar is bench mined. The net effect is that partially benched pillars experience a simultaneous reduction in strength and load.
6. The instability of the partly benched pillar can be further ascribed to the uneven distribution of stresses within the pillars when they are located at the edge of a benching operation. High local stresses near the top and bottom of the pillar can initiate stress spalling.

For pillar design purposes, it appears that pillars that will be benched should be designed to accommodate an increase of about 15% in the average stress while they are in the prebenched state. However, pillars that are designed to be stable at the maximum benched height should also be stable under the induced stress increase before benching, owing to the greater strength of the shorter pillars before benching. No special design modification is, therefore, necessary for pillars that will be bench mined.

Pillar Length Effect on Pillar Strength

This investigation was conducted to determine the degree to which pillar strength can be increased by using rectangular pillars over the more standard square pillars. Square pillars are widely used in stone mines, but rectangular pillars have been used in situations where horizontal stress is an issue. Through the use of rectangular pillars, the roof exposed in the direction of the maximum horizontal stress can be minimized [Iannacchione et al. 2003]. Rectangular pillars can also increase the efficiency of ventilation in the stone mines [Grau et al. 2006]. With longer pillars, the number of ventilation stoppings can be reduced.

There have been a number of equations developed to predict the increase in strength from square to rectangular pillars. Many of these equations have not been substantiated or have been used generically for coal mine pillars which are not as slender as pillars used in stone mines. Another consideration for stone mines is the type of failure that can occur in high openings with slender pillars. Pillar spalling and brittle rock failure occur at stress levels well below the expected rock and pillar strength. Brittle failure occurs when the confinement of the rock is low; it is not clear whether slender pillars would experience the same benefit from increased length as would wider pillars.

Numerical models that simulated the brittle failure process were used to evaluate the effect of pillar length on the pillar strength. Details of the model setup, input parameters, and failure criteria were presented by Dolinar and Esterhuizen [2007]. The models were evaluated for various width-to-height ratios and length-to-width ratios. Figure 24 shows numerical model results for the strength increase against pillar length, expressed as a ratio of the rectangular to square pillar strength. There is a large difference in the gain in pillar strength with length depending on the width-to-height ratio of the pillar. An increase in strength of over 40% occurred for the squattest pillar model (i.e., largest width-to-height ratio). The results showed that, for a width-to-height ratio of less than 0.6, there is little or no increase in strength with increased length. As the width-to-height ratio increases, the benefits of increased length are greater. This result appears to be related to the absence of confinement in the more slender pillars.

Some of the accepted methods of estimating the effect of length on pillar strength do not account for the fact that the strength increase might be limited when pillars are slender. The methods would generally predict a similar strength increase for all pillars, regardless of the width-to-height ratio. Therefore, when designing slender pillars for brittle rock, it is necessary to consider the width-to-height ratio as well as the length-to-width ratio when calculating the strength of rectangular pillars.

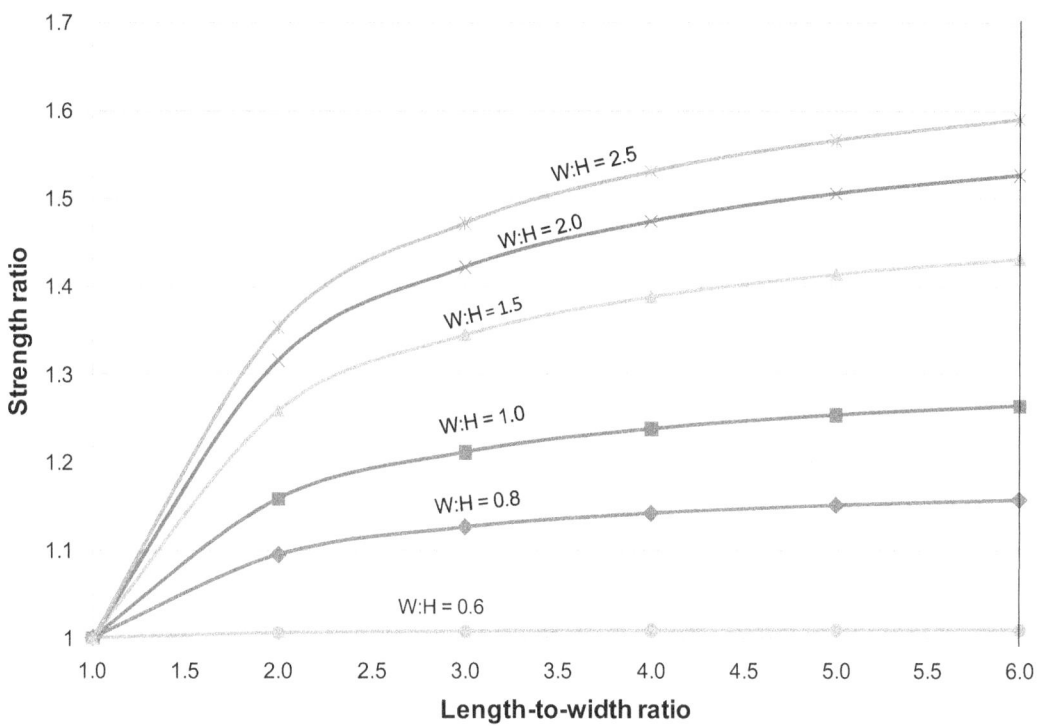

Figure 24. Strength increase caused by increasing pillar length for pillars with various width-to-height ratios. Results from calibrated numerical models assuming rock failure initiates as spalling followed by shearing.

Pillar Strength Equation for Stone Mines

To estimate the strength of pillars in stone mines, an equation was developed by combining empirical field data and analytical results with information from other mining operations that are similar to stone mine room-and-pillar workings. The database on stone mine pillar performance contains information on many stable pillar systems but only 18 individual failed pillars, which are likely to be the weakest members of the population of pillars. These data are, therefore, not representative of the average stone pillar strength and are not sufficient to develop a purely empirical strength equation for stone mines. For this reason it was necessary to expand the investigation to include the results of numerical models and data from other mining operations.

Base Equation

Records of stable and failed pillars in the lead mines of the Viburnum Trend in Southeastern Missouri were considered to be the most appropriate for developing a strength equation for stone mines. The workings are flat-lying and room-and-pillar operations have been conducted with mostly square pillars [Carmack et al. 2001] since the 1960s. The host rock is dolomitized limestone with strength characteristics similar to the limestone generally mined in stone operations. The average UCS of the rock is approximately 22,000 psi [Roberts 2005], which falls within the upper range of limestone formation strengths. The rock mass quality was assessed at several different underground locations by the authors and found to fall within the range found in stone mines. It is recognized that the presence of mineralization within the host rock can affect rock failure mechanisms and post-failure behavior. However, the stages of failure development, observed underground and reported by Lane et al. [1999] are very similar to those seen in stone mine pillars. Importantly, a wide-area pillar collapse occurred at one of the mine operations during the 1980s and the details of the pillar dimensions and mode of failure were investigated in detail, which provides valuable data on the ultimate pillar strength [Zipf 2001].

A well-documented pillar design procedure has been developed for these mines based on the observation of failed and stable pillars [Lane et al. 1999; Roberts et al. 2007]. The design technique makes use of numerical models to estimate pillar loading while pillar strength is estimated by a set of strength relationships which are based on the confinement principle, based on the approach of Lunder [1994]. Direct observations of hundreds of pillars, which included both stable and failed case histories, have been used to refine the strength relationships.

In principle, the pillar strength is determined by viewing each pillar in plan and subdividing the pillar into 2.4 m (8 ft) square elements. Each element is labeled as an "outer" or "inner" element. The outer elements have a lower strength than the inner elements owing to the lack of confinement. The strength is also affected by the pillar height, according to relationships presented in Roberts et al. [2007]. For example, a 4.8 m (16 ft) square pillar will consist of four 2.4 m (8 ft) "outer" elements and will be weaker than a 7.2 m (24 ft) square pillar that has eight "outer" and one "inner" element. The method, therefore, takes into consideration both the pillar shape and pillar volume for estimating pillar strength.

In order to express the pillar strength relationships in the form of a power equation, a series of strength curves were developed for various pillar widths using the "inner" and "outer" element approach. The parameters for a power equation were then determined by the least squares curve-fitting technique. The following equation was obtained:

$$S = k \times \frac{w^{0.30}}{h^{0.59}} \quad (3)$$

where w is pillar width and h is pillar height. The strength parameter k was found to be 140 MPa (20,300 psi). The value of k can be expressed in terms of the UCS when using dimensions in meters as shown next:

$$k = 0.65 \times UCS \quad (4)$$

based on the average UCS value of 152 MPa (22,000 psi) for the formation. Note that, for pillar dimensions in feet, the k parameter becomes $0.92 \times$ UCS.

Adjustment for the Presence of Large Discontinuities

The field data and analysis, presented in this document, shows the necessity of accounting for the impact of large, angular discontinuities on pillar strength. Such an adjustment should include both the inclination and spacing of the large discontinuities. Large discontinuities can be widely spaced and do not necessarily intersect each pillar in a layout. The results of numerical models, discussed in this document, were used to develop adjustment factors for large, angular discontinuities. Table 5 lists discontinuity dip factors (DDF) that relate to the strength of pillars intersected by single, large discontinuities to the undisturbed pillar strength; these DDFs directly relate to the numerical model results shown in Figure 16. Table 5 shows that discontinuities that dip at about 30° to 70° can have a significant impact on pillar strength, and the impact is exacerbated as the width-to-height ratio decreases.

Table 5. Discontinuity dip factor (DDF) representing the strength reduction caused by a single discontinuity intersecting a pillar at or near its center, used in equation 5.

Discontinuity dip (°)	Pillar width-to-height ratio								
	≤0.5	0.6	0.7	0.8	0.9	1.0	1.1	1.2	>1.2
30°	0.15	0.15	0.15	0.15	0.16	0.16	0.16	0.16	0.16
40°	0.23	0.26	0.27	0.27	0.25	0.24	0.23	0.23	0.22
50°	0.61	0.65	0.61	0.53	0.44	0.37	0.33	0.30	0.28
60°	0.94	0.86	0.72	0.56	0.43	0.34	0.29	0.26	0.24
70°	0.83	0.68	0.52	0.39	0.30	0.24	0.21	0.20	0.18
80°	0.53	0.41	0.31	0.25	0.20	0.18	0.17	0.16	0.16
90°	0.31	0.25	0.21	0.18	0.17	0.16	0.16	0.15	0.15

The DDF values shown in Table 5 are applicable for use when considering the stability of a single pillar that is intersected by a large discontinuity. However, these values would be conservative for assessing a layout of many pillars, because large discontinuities can be widely spaced and may not necessarily intersect every pillar. The average impact of large discontinuities on the strength of pillars in a layout is referred to as the large discontinuity factor (LDF), and it can be estimated as shown in the following equation:

$$LDF = 1 - DDF \times FF \qquad (5)$$

where DDF is the discontinuity dip factor shown in Table 5, and FF is a frequency factor related to the frequency of large discontinuities per pillar shown in Table 6.

Table 6. Frequency factor (FF) used in equation 5 to account for large discontinuities.

Average frequency of large discontinuities per pillar	0.0	0.1	0.2	0.3	0.5	1.0	2.0	3.0	>3.0
Frequency factor (FF)	0.00	0.10	0.18	0.26	0.39	0.63	0.86	0.95	1.00

If there are no large discontinuities present, the FF is equal to zero and the LDF will equal 1.0, having no effect on pillar strength. The frequency of large discontinuities per pillar can be estimated easily by dividing the pillar width by the average spacing of the large discontinuities. For example, if we are designing pillars that are 9 m (30 ft) wide at a width-to-height ratio of 1.0

and we want to know the impact of large discontinuities that are spaced 30 m (100 ft) apart, dipping at 50°, we can proceed as follows:

1. Calculate the expected frequency of large discontinuities per pillar (9/30 = 0.3),
2. In Table 6 we find that FF = 0.26
3. Look up the discontinuity impact factor (DDF) in Table 5, which is 0.37.
4. Calculate the LDF for the pillar layout using equation 5, which is 0.90.

The LDF value of 0.9 represents a 10% reduction in the average strength of pillars in the layout. Using a reduced average strength to design a pillar layout will ensure that the layout as a whole is stable. However, the individual pillars that are intersected by large, angular discontinuities may become unstable when they are formed. This potential instability and the required remedial actions to ensure safe mining operations near an affected pillar should be considered during the design stage.

The field observations of pillar performance did not include any cases where every pillar in a layout was intersected by one or more angular (30° to 70° dip) discontinuities. Therefore, the validity of the LDF under such conditions could not be verified. It is recommended that a detailed rock engineering investigation and pillar strength assessment should be conducted if more than about 30% of the pillars are expected to be intersected by large discontinuities that dip between 30° and 70°.

Adjustment for Rectangular Pillars

The analyses described in this document indicated that slender pillars in brittle rock do not benefit as much from a length increase as wider pillars. This is caused by the lack of confinement in the slender pillars. The numerical model results indicate that the length benefit is likely to be zero when a pillar has a width-to-height ratio of 0.5 and it gradually increases as the width-to-height ratio approaches 1.4, when the full length benefit is realized. A length benefit ratio (LBR) is introduced to account for the width-to-height ratio effect of slender pillars in brittle rock. The LBR is zero when the width-to-height ratio is 0.5 and gradually increases to 1.0 at a width-to-height ratio of 1.4, when the full length benefit is realized. A similar approach has been used to estimate the strength of rectangular pillars in Australian coal mines [Galvin et al. 1999].

The "equivalent width method", proposed by Wagner [1992], was used as a basis for calculating the length benefit. According to this method, the strength increase of a rectangular pillar is expressed as an equivalent increase in pillar width, which then replaces the true pillar width in the pillar strength equation. A modified form of Wagner's equivalent width equation, which includes the LBR, is proposed as follows:

$$w_e = w + \left(\frac{4A}{C} - w\right) \times LBR \tag{6}$$

where w is the minimum width of the pillar, A is the pillar plan area, C is the circumference of the pillar, and LBR is the length benefit ratio. Table 7 shows the suggested relationship between width-to-height ratio and the value of LBR based on the modeling results presented in section 5.5. Using this approach, the value of w_e will equal the pillar width w when pillars are square.

Table 7. Values of the length benefit ratio (LBR) for rectangular pillars with various width-to-height ratios.

Width-to-height ratio	0.5	0.6	0.7	0.8	0.9	1.0	1.1	1.2	1.3	1.4
Length benefit ratio (LBR)	0.00	0.06	0.22	0.50	0.76	0.89	0.96	0.98	0.99	1.00

Adjustments for Other Geotechnical Conditions

Observations showed that the presence of thin weak bedding bands appears to have contributed to the failure of several of the pillars presented in Table 3. The data and understanding of this failure mode are not sufficient to account for weak bands in the pillar strength equation. The information on the impact of weak floor strata on stone mine pillar performance is similarly limited. The authors suggest that the services of a rock engineering specialist should be sought when these conditions exist, so that a detailed program of investigation can be conducted.

Pillar Strength Equation Modified for Stone Mines

The base equation for stone mine pillar strength (equation 3) can be written as follows to include the intact rock strength and the adjustment for large discontinuities:

$$S = 0.65 \times UCS \times LDF \times \frac{w^{0.30}}{h^{0.59}} \tag{7}$$

where UCS is the uniaxial compressive strength of the intact rock, LDF is the large discontinuity factor, w and h are the pillar width and height in meters. When using dimensions in feet, the 0.65 constant becomes 0.92. For rectangular pillars w is replaced by the equivalent width w_e, which can be calculated using equation 6. The value of LDF can be determined from equation 5. If no large discontinuities are present, the LDF will equal 1.0.

Pillar Factor of Safety Determination

Equation 7 was used to calculate the adjusted strength and FOS of all the pillars in the stone mine database. The results are presented in Figure 25, which displays the calculated FOS against the width-to-height ratio. Various symbols were used to indicate currently operating and disused layouts, failed pillars, and the approximate number of pillars in the various layouts. Disused layouts may have been abandoned because of stability concerns or changes in operating procedures. However, all the disused and current layouts are considered to be "successful" because they were all successfully supporting the overlying strata at the time of this study. The FOS axis shows values up to 10.0; this means that 13 cases with FOS values greater than 10.0 are not shown.

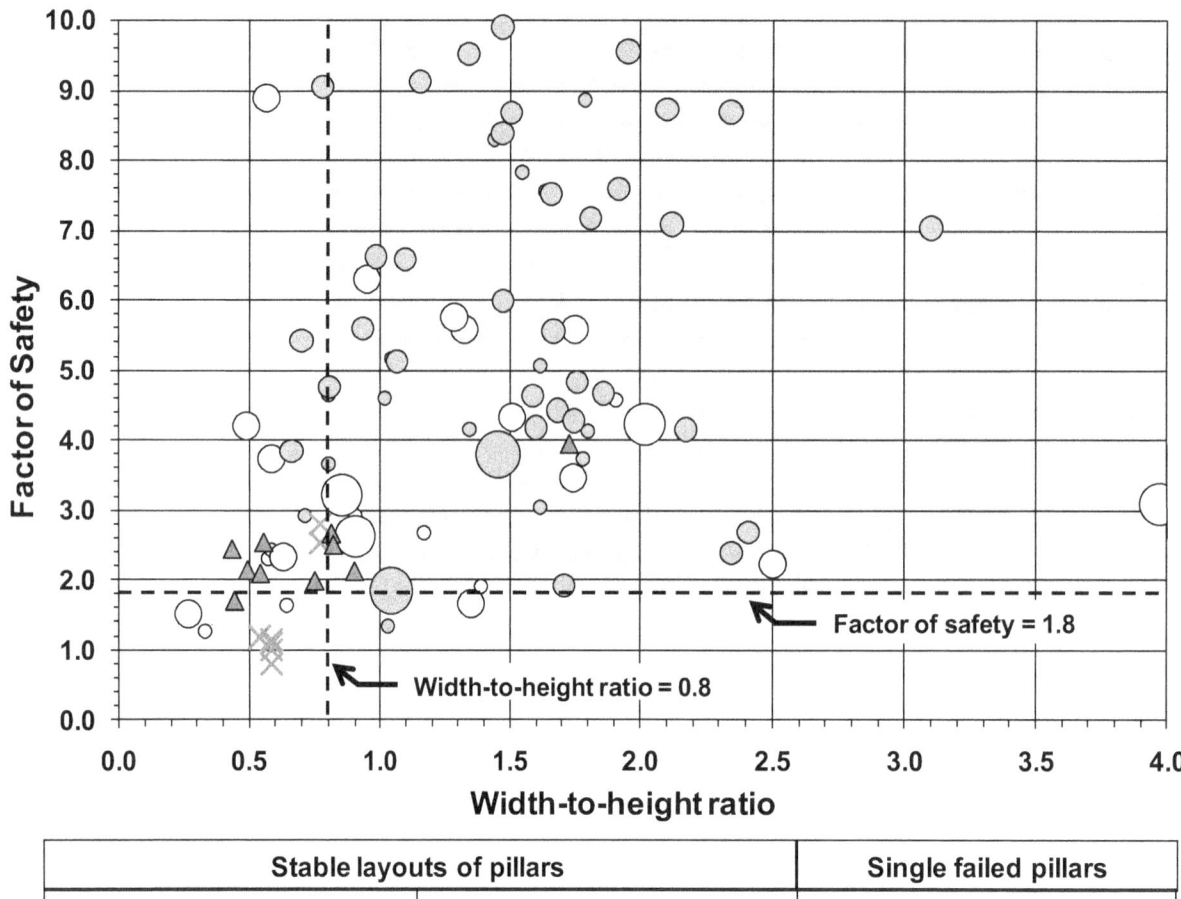

Figure 25. Chart showing the factor of safety against width-to-height ratio using equation 7. Current and disused pillar layouts are shown as well as single failed pillars. The recommended area for pillar design is shaded.

The calculated average FOS of all the failed pillars is 2.0, which includes the cases that were intersected by large angular discontinuities. The calculated average FOS of pillars that are intersected by large discontinuities is 1.5. The minimum FOS for the successful layouts was calculated as 1.27, which is one of the disused layouts.

Applicability of the Pillar Design Equation

The factor of safety chart shows that equation 7 provides reasonable results for the observed pillars in operating mines. The failed pillars are seen to have a lower calculated FOS than most of the stable pillars and the stable pillars all have FOS greater than 1.0. The equation is strongly based on empirical observations and should, therefore, not be used in cases that fall outside the limits of the case histories.

Roof Span Design Considerations

Background

In room-and-pillar mines, the roof between the pillars is required to remain stable during mining operations for haulage as well as access to the working areas. In underground stone mines, the size of the rooms is largely dictated by the size of the mining equipment. Underground stone mines use large mining equipment to operate economically and require openings that are on average 13.5-m (44-ft) wide by approximately 7.5-m (25-ft) high to operate effectively. The desired roof span dimensions are largely predetermined by the operational requirements and design is focused on optimizing stability under the prevailing rock conditions. If the rock mass conditions are such that the desired stable spans cannot be achieved cost effectively, it is unlikely that underground mining will proceed. NIOSH research into stone mine roof stability has focused, therefore, on identifying the causes of instability and techniques to optimize stability through design.

Methods of Roof Span Design

Designing stable roof spans in underground mining excavations is largely conducted using empirically based techniques and may be supplemented by analytical or computational methods. Empirical methods based on rock mass classification [Bieniawski 1989; Barton et al. 1974; Mathews et al. 1980; Laubscher 1990] are widely used to obtain an initial indication of likely stable spans that can be achieved under given rock mass conditions. The classifications can also be used to obtain an estimate of the support requirements.

In some cases analytical equations based on elastic theory can be used to calculate likely roof deflection and stresses, but these methods have found limited application owing to the complex behavior of rock which contains many defects and variable properties that do not follow the assumptions of isotropic, linear elastic materials. However, useful insight into the parameters which affect roof stability under idealized conditions can be obtained. Owing to these shortcomings, numerical models that can simulate the effect of discontinuities in rock and model the nonlinear rock mass response after it has failed have found wide acceptance as design tools. However, there are no universally accepted methods to assess the safety of an excavation or the acceptability of a design [Hoek et al. 2008]; engineering judgment and experience continues to play a large role in the design layout. Therefore, a pragmatic approach is proposed for stone mine roof span design, in which the designer systematically considers the rock mass characteristics and stress conditions to develop a mine layout and select the excavation span. The method relies heavily on information collected on the past performance of stone mine workings in the Midwestern and Eastern United States.

Stability of Bedded Rock

The mining operations included in this study were all mining bedded stone deposits. The presence of bedding within the rock mass causes some unique advantages and stability issues not present in many other mineral deposits. In bedded rocks the bedding planes subdivide the rock into plates of varying thickness which will bend and deflect into the mine openings. If the span of the opening is too wide, the bed deflection can become excessive which can result in tensile cracking near the center of the roof span or crushing along the edges. Excessive downward deflection of the roof can also result in loosening of blocks of rock that are defined by high angle joints in the rock, and can lead to the collapse of the roof. Thin roof beds will obviously deflect more than thicker beds.

The presence of any other discontinuities, such as steeply dipping joints or faults, also needs to be considered. A continuous steeply dipping discontinuity will destroy the continuity of the beams in the roof that can result in local instability. Multiple intersecting joints can create a blocky roof condition that is difficult to support.

The presence of bedding planes in the rock mass can be advantageous to roof stability if the roofline coincides with a well-developed bedding plane. The bedding plane helps to limit blast damage to the roof and presents a clean breaking surface for blasting operations.

When designing excavations in bedded rock, therefore, it is necessary to develop a clear understanding of the nature of the bedding planes in the rock mass and to determine whether they are likely to remain stable over the proposed excavation spans.

Developing Roof Span Design Guidelines for Stone Mines

The roof span design procedures presented in this document were developed following a combined empirical and analytical approach. The actual performance of the roof in 34 different stone mines in the Midwestern and Eastern United States were recorded. At each mine data on rock mass conditions, discontinuities, roof span dimensions, support methods, and factors that contributed to instability were recorded. Supplemental data on defects within the roof were collected using a borehole video camera at 13 different mines; roof monitoring data from 15 different mine operations were considered. The field data was evaluated in terms of existing rock classification systems and expected roof span stability. Issues, such as the impact of horizontal stress on roof stability, were further investigated using numerical models. The results of these studies were evaluated and form the basis for the design procedure that follows.

Survey of Roof Span Performance

A survey of the roof stability conditions was made at a total of 92 locations in 34 different stone mines. The basic rock mass data and intact rock strength information are summarized in the Geotechnical Characteristics section of this document. In addition, measurements were made of the room width and the diagonal span across four-way intersections. Table 2 provides a summary of the excavation and pillar dimensions, showing that room widths vary from 9.1 m (30 ft) to 16.8 m (55 ft). Figure 26 shows 63 of the 92 observed room widths fall in the 12.2–15.2 m (40–50 ft) range. The diagonal span measured at four-way intersections averaged 21.7 m (70 ft).

A total of 54% of the sites investigated were naturally stable and did not need regular roof support. Some of these mines occasionally used roof bolts to support the roof in isolated areas. Figure 27 shows an example of a 13.4-m (44-ft) wide, naturally stable excavation with excellent roof conditions. Regular reinforcement by pattern bolting or irregularly spaced bolts was observed at the remaining 46% of the locations visited.

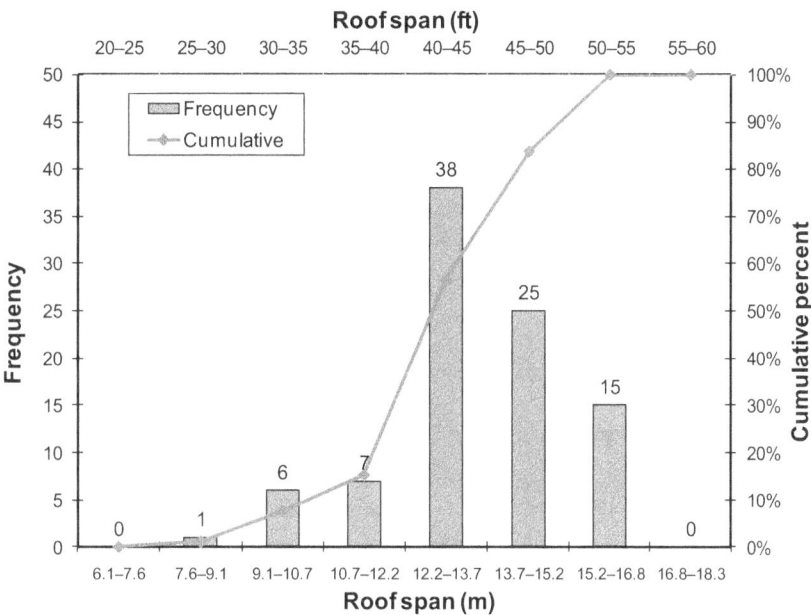

Figure 26. Distribution of roof span dimensions measured at 34 different underground stone mines.

Figure 27. Naturally stable 13.4-m (44-ft) wide roof span in a stone mine.

Roof Instabilities

All but 4 of the 34 mines visited had experienced some form of roof instability. Small scale roof falls were observed that typically consist of single rock fragments that are less than 30 cm (1 ft) across; larger roof falls typically consist of one or more rock fragments that are larger than 30 cm (1ft) across and can extend over the full width of an excavation. The instability factors present in small-scale roof falls were categorized as follows:

- Thin slivers of rock that did not appear to be bounded by natural joints. They may be related to blast damage, stress spalling, or time-related weathering.
- Blocks that were defined by intersecting joint planes and bedding planes.
- Beams or plates that were formed by bedding planes.

The small-scale rock falls affected about 28% of the total roof area that was evaluated. In the remaining areas the roof was stable with no sign of current or past instability. Most of the above-listed instabilities can be addressed by scaling, rock bolting, or screen installation as part of the normal support and rehabilitation activities.

In addition to small-scale rock falls, large falls were observed at 19 mine operations. The large falls made up a very small percentage of the exposed roof in the mines; many of the mines only had a single instance of a large roof fall. The large roof falls were categorized by identifying the most significant factor that appeared to contribute to each fall. A summary of these factors and the relative frequency of occurrence of each are presented below:

- **Horizontal stress.** High horizontal stress was assessed to be the main contributing factor in 36% of all roof falls observed. These falls appeared to be equally likely to occur in shallow or deep cover. A roof fall related to stress-induced damage was observed in one case at a depth of as little as 45 m (150 ft).
- **Bedding-related roof beams.** The beam of rock between the roofline and an overlying weak band or parting plane failed in 28% of all roof falls observed.
- **Blocks defined by large discontinuities.** Large discontinuities extending across the full width of a room contributed to 21% of the roof falls.
- **Caving of weak overlying strata.** The remaining 15% of the roof falls was attributed to the collapse of weak shale or progressive failure of low-strength roof rocks.

Although the large roof falls only make up a small percentage of the total roof exposure, their potential impact on safety and mine operations can be very significant. Most cases of large roof falls required barricading off or abandonment of the affected entry. When large roof falls occur in critical excavation areas, the repair can be very costly. Figure 28 shows a case where extensive support was required to rehabilitate a large roof fall.

Figure 28. Bolts, straps, and injection grouting used to rehabilitate the roof at the site of a major roof fall.

Support Practices

The survey of roof support practices showed that grouted rock bolts are the most widely used form of support. Rock bolts of various types are used to reinforce the roof. Fully-grouted bolts are the most commonly used bolts; friction bolts and mechanical anchor bolts are also used, but are less prevalent. Bolt lengths vary from 0.9 m (3 ft) to 2.4 m (8 ft) with 1.8-m and 2.4-m (6-ft and 8-ft) long bolts making up 67% of the bolts included in the survey. Bolt spacing of 1.5 m (5 ft) and 1.8 m (6 ft) are the two most commonly observed spacings, and the maximum bolt spacing was 2.4 m (8 ft). As with most other roof bolting designs in strong rocks, high strength and stiff bolts are more likely to provide the desired rock reinforcement than low strength and low stiffness systems [Iannacchione et al. 1998].

In extreme situations cable bolts and sealant injection have been used to stabilize the roof; but roof screen is rarely used. These items are considered special applications and were not included in the study.

Comparison of Roof Stability in the Physiographic Regions

An evaluation was made of the collected data to determine whether differences exist in rock conditions, and in roof stability when comparing the Appalachian Highlands and Interior Plains physiographic regions. The evaluation showed that the regions are very similar in terms of rock mass strength as expressed by the RMR values. The average uniaxial compressive strength of the rocks in the Appalachian Highlands region appears to be slightly higher, but insufficient data is available to determine the level of statistical significance. The average room width in the Appalachian Highlands is 13.7 m (45.0 ft); in the Interior Plains the average room width is 13.5 m (44.2 ft). These similar average room widths indicate that the rock conditions in both regions are likely to be similar, allowing similar excavation dimensions to be developed. Roof bolting is used in about 50% of the mines in both regions, again confirming that rock conditions are similar. Roof bolt spacing and lengths were not significantly different in the two regions. Horizontal stress-related roof stability issues were also equally prevalent in the two regions.

Stone Mine Roof Stability Analysis

Roof Span Dimensions

Roof span size is closely related to a mine's capacity to effectively operate large loaders and haul trucks. The majority of roof spans in operating mines fall within a narrow range of 9.1 m to 16.8 m (30 ft to 55 ft), which is generally sufficient space to effectively operate this equipment. Few of the mines used roof spans wider than 15 m (50 ft), so it is not clear whether the stability limit is approached when the heading width exceeds 15 m (50 ft) or whether it simply satisfies the practical requirements for equipment operation. Given that a large proportion of the mines are able to mine without installed support, it seems to indicate that wider spans can be achieved if additional supports are used. Whether these larger spans would be cost effective will, of course, depend on the support costs.

One way of assessing the potential maximum span is to compare the stone mine data to experience in other mine openings around the world. The Stability Chart originally developed by Mathews et al. [1980], then modified by Potvin [1988], Nickson [1992], and Hutchinson and Diederichs [1996], was used as a basis for comparison. The Stability Chart plots a modified stability number N' which represents the rock mass quality normalized by a stress factor, an orientation factor and a gravity adjustment. Figure 29 shows four different stability zones that have been developed, based on 176 case histories from hard rock mines around the world. In this chart the actual heading width is shown instead of the hydraulic radius, which is customarily used. The conversion from hydraulic radius to heading width assumes the heading is a parallel-sided excavation. The increased effective width associated with intersections is implied in the stone mine case histories because the data includes both intersections and parallel-sided heading failures. The four stability zones in the Stability Chart are as follows:

- **Stable.** Support generally not required.
- **Stable with support.** Support required for stability; the support type is cable bolting.
- **Transition zone.** Stability not guaranteed, even with cable bolt support.
- **Unsupportable.** Caving occurs; cannot be supported with cable bolts.

The stability number was calculated for each of the 92 stone mine sites and plotted on the Stability Chart shown in Figure 29. The chart also indicates the average stability number for stone mines as a horizontal dashed line.

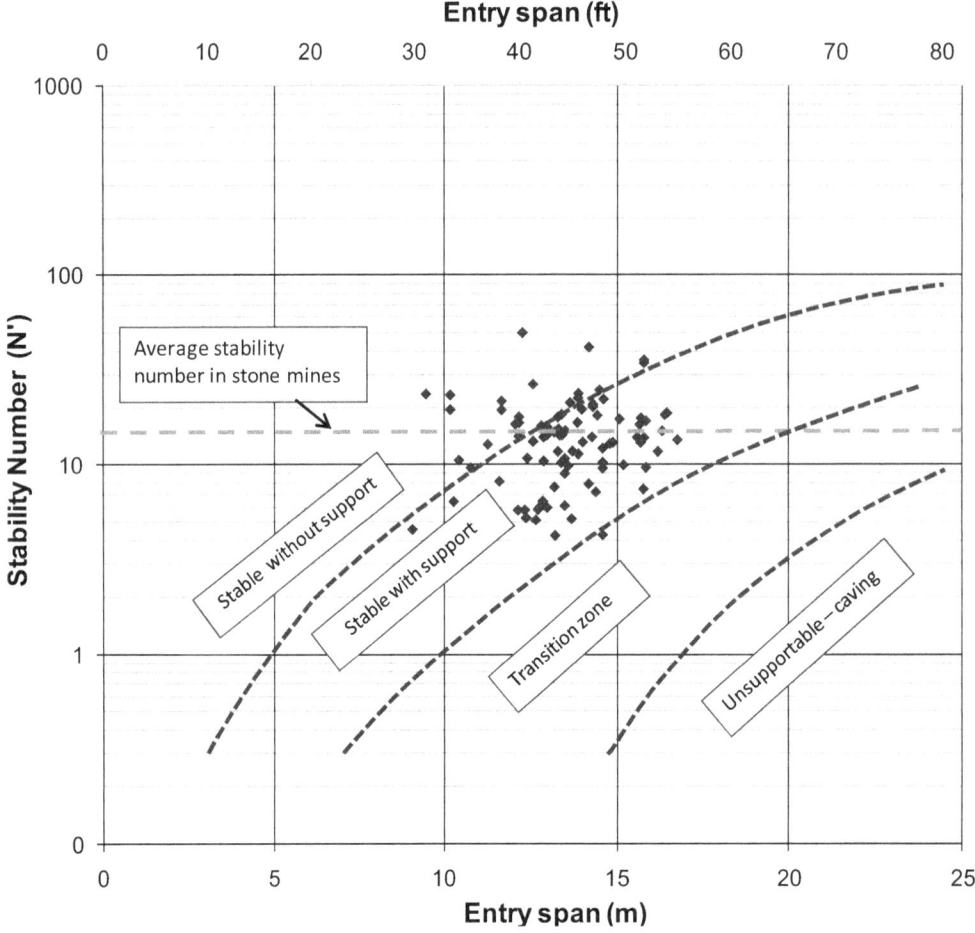

Figure 29. Stability chart showing stone mine case histories and stability zones, modified after Mathews et al. [1980], Potvin [1988], Nickson [1992], and Hutchinson and Diederichs [1996].

Figure 29 shows that the majority of stone mine case histories plot in the region of "stable" to "stable with support" and only one is located in the transition zone. This agrees reasonably well with the observed stability and support present in stone mines, although stone mines have been able to achieve stability with light support compared to cable bolting used in the hard rock mine case histories. Based on the average stability number for stone mines, it would appear that stable, supported excavations can reliably be achieved with spans of up to about 20 m (65 ft) using cable bolt supports. Unsupportable conditions are likely when the span increases to about 27 m (90 ft). These results are in line with current experience. It appears that stone mines are working near the span limit that can be reliably achieved using rock bolts as the support system. Increasing the spans beyond the 15–17 m range (50–55 ft) is likely to incur considerable cost and productivity implications as cable bolting would become necessary.

Thickness of the Immediate Roof Beam

The stability of excavations in bedded deposits is closely tied to the composition and thickness of the first beam of rock in the roof. An assessment of the data collected showed that 25 of 34 mines were attempting to maintain a specific thickness of rock in the immediate roof. In some cases the upper surface of the beam was a pronounced parting plane; in others, it was a change in lithology, typically when the rock beam is overlaid by weaker materials. A constant thickness of roof beam is achieved either by probe drilling to determine the thickness of the roof beam or by following a known parting plane or marker horizon.

Several of the mines that used regular support did so to alleviate the effects of horizontal stress, which is not related to beam thickness. If these mines are removed from the data, the average roof beam thickness in mines that use regular support is 0.7 m (2.3 ft). Figure 30 shows the effect of the roof beam thickness on excavation stability in mines that did not experience horizontal stress related instability. It can be seen that when the beam is equal to or less than 1.2 m (4 ft), support is likely to be required to maintain stability, or the excavation may be unstable. Of the locations where the roof beam thickness was 1.2 m (4 ft) or less, 82% were unstable or required support to maintain stability. These results seem to indicate that mines with a relatively thin beam of rock in the immediate roof are more likely to encounter an unstable roof, and regular roof bolting becomes necessary. There was no correlation between roof beam thickness and roof span.

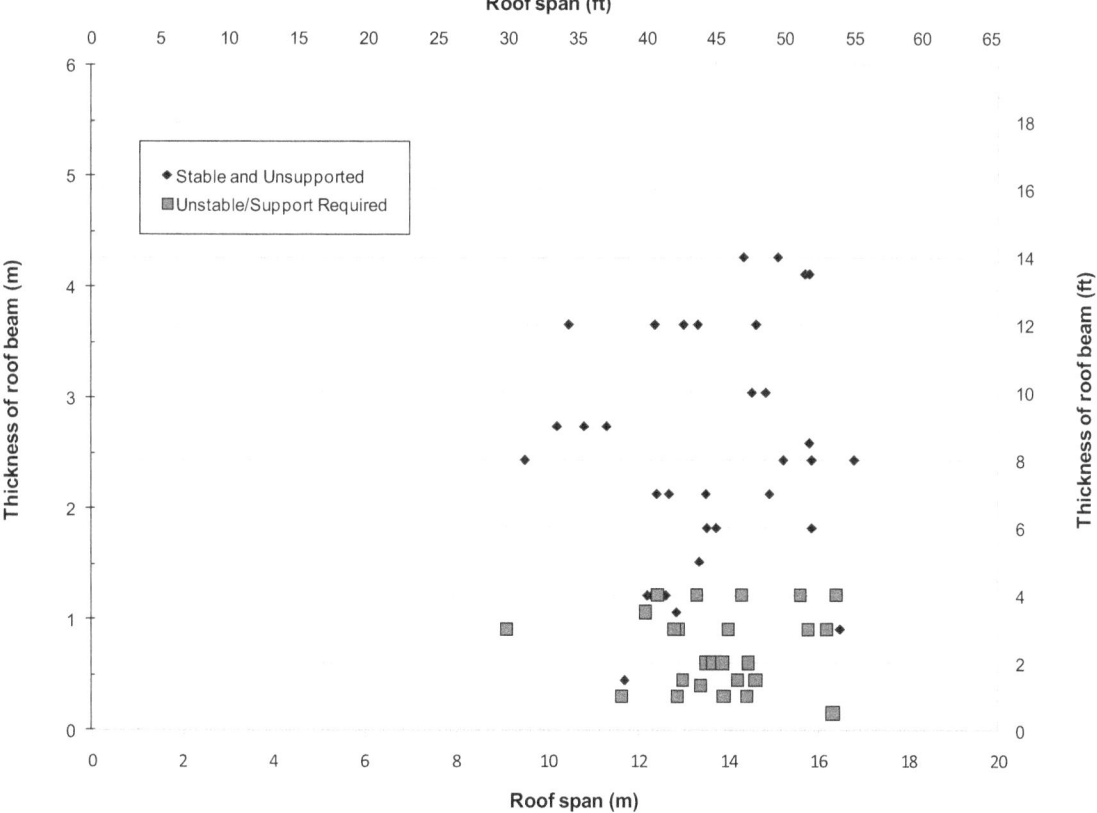

Figure 30. Chart showing the effect of the thickness of the roof beam on excavation stability in mines that did not experience horizontal stress related instability.

The beam thickness is obviously not the only factor to consider when deciding on roof reinforcement. Other aspects such as roof jointing, bedding breaks, blast damage, groundwater, and horizontal stress can contribute to roof instability resulting in the need for rock bolt support. However, the experience seems to indicate that a roof beam of less than about 1.2 m (4 ft) is highly likely to be unstable, and a regular pattern of rock bolt supports will be required to maintain the roof stability.

Horizontal Stress Issues

Horizontal stress can cause beams within the roof to buckle and fail in shear [Iannacchione et al. 2003]. Failure can initiate as guttering in one corner of an excavation, (i.e., called "cutter roof" in coal mines), as shown in Figure 31, and can propagate to a large-scale roof fall, as shown in Figure 32. Falls related to horizontal stress typically line up in the direction perpendicular to the regional maximum horizontal stress and are oval shaped when seen in plan view, as shown in Figure 33. The falls tend to propagate laterally in the direction perpendicular to the main horizontal stress, and can snake through the mine, as shown in Figure 34. Careful observation of the roof falls, their direction of propagation, and other signs of excessive stress can assist in identifying the orientation of the maximum horizontal stress [Mark and Mucho 1994].

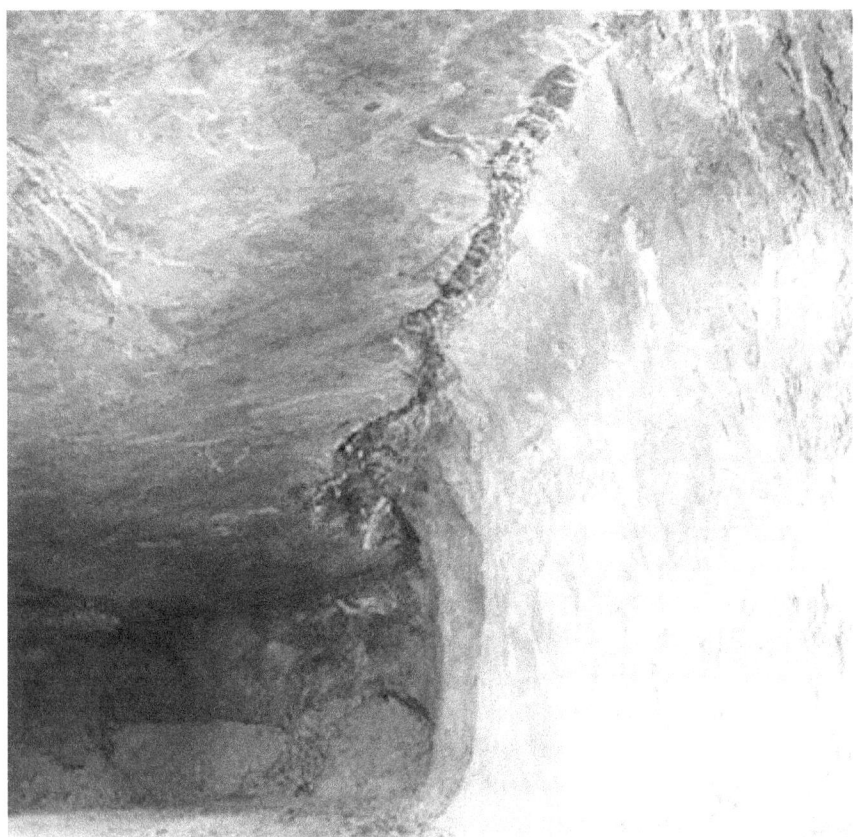

Figure 31. Roof guttering at the pillar-roof contact.

Figure 32. Large stress-related, oval-shaped fall that has propagated upwards into weaker, overlying strata in a limestone mine.

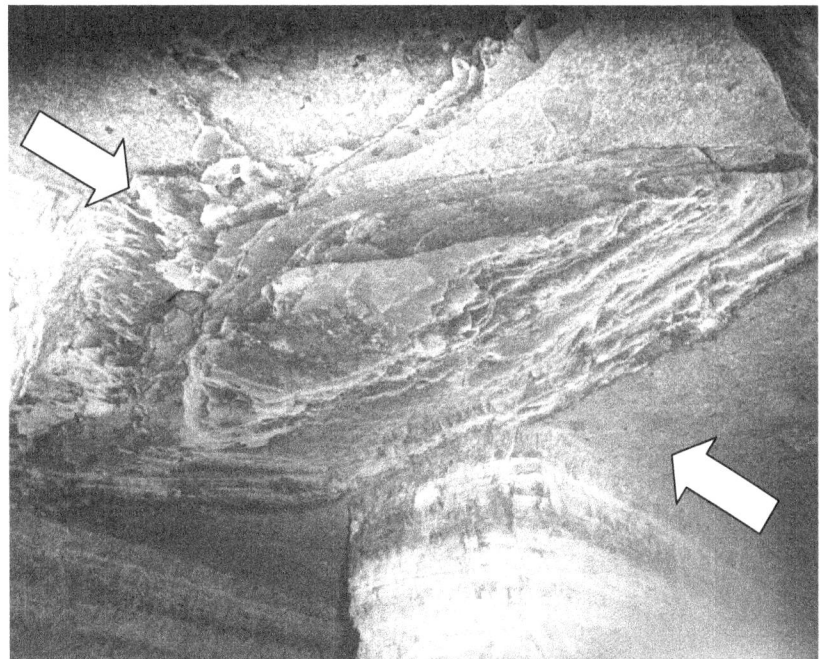

Figure 33. Horizontal, stress-induced roof failure that initiated between two pillars. Arrows show measured direction of maximum horizontal stress.

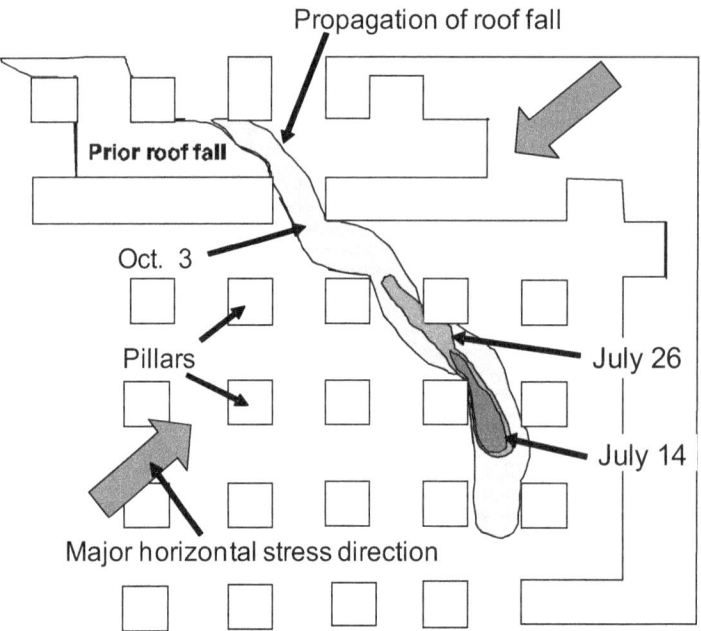

Figure 34. Plan view showing the development of a stress-related roof fall in the direction perpendicular to the direction of the major horizontal stress [Iannacchione et al. 2003].

The field results showed that horizontal, stress-related roof instability can occur at any depth of cover [Esterhuizen et al. 2007]. This is not unexpected, given that the horizontal stress is caused by tectonic compression of the limestone layers, which is not related to the depth of typical stone mines [Dolinar 2003; Iannacchione et al. 2003].

An analysis of the impact of horizontal stress on bedding-defined beams of rock that may exist in the roof of stone mine workings showed that horizontal stress can cause elastic buckling of thinly bedded roof strata [Iannacchione et al. 1998]. Elastic buckling can lead to failure of brittle rock, particularly when the tensile strength of the rock is exceeded.

Horizontal, stress-driven roof failures also occur in roof rocks that are not necessarily thinly bedded. In these cases, the failure can be explained by considering the brittle spalling mechanism of failure, which is often observed in pillar ribs. This failure mode can occur at stress magnitudes that are much lower than the intact rock strength. Figure 35 shows a curved extension fracture in the rib of a long, rectangular pillar that was exposed when a cross-cut was developed through the pillar several years after the initial development of the pillar. The average stress in this pillar is estimated to be in the range of 15 to 20 MPa (2,200 to 2,900 psi), which is similar to the horizontal stress that can be expected to exist in the roof of stone mines. It is, therefore, likely that similar extension fractures can be expected to exist in the roof of stone mine excavations.

Numerical models were used to investigate how this type of failure might take place and how the presence of widely spaced bedding planes would affect the depth of the potential roof failure [Esterhuizen 2006]. The stability of the roof rocks was assessed by calculating a failure index, which is based on extension failure initiating when the maximum principal stress exceeds 10% of the rock strength. A failure index of less than 1.0 indicates potential rock failure, similar to the traditional factor of safety. The failure index results in Figure 36a show that, in the absence of bedding discontinuities, extension fracturing can extend over the room to form an arch which extends to about 3 m (10 ft) above the roofline. This arch is very similar to the extension fracture observed in the pillar rib, shown in Figure 35.

Figure 35. Extension fractures exposed in a pillar rib after the pillar was bisected by a new crosscut.

If a single bedding discontinuity is introduced 1 m (3.3 ft) above the roofline, see Figure 36b, the stresses are redistributed by the presence of the discontinuity. A reduction of the horizontal stress occurs in the 1-m (3.3-ft) thick roof beam as it deflects downwards, and some slip occurs along the bedding discontinuities. Separation of up to 2 mm (0.08 in) occurs across the bedding discontinuity near the center of the room. The deflection of the lower beam causes an increase in the stress within the overlying roof, which in turn causes the potential rock failure to extend to about 4 m (13 ft) above the roofline.

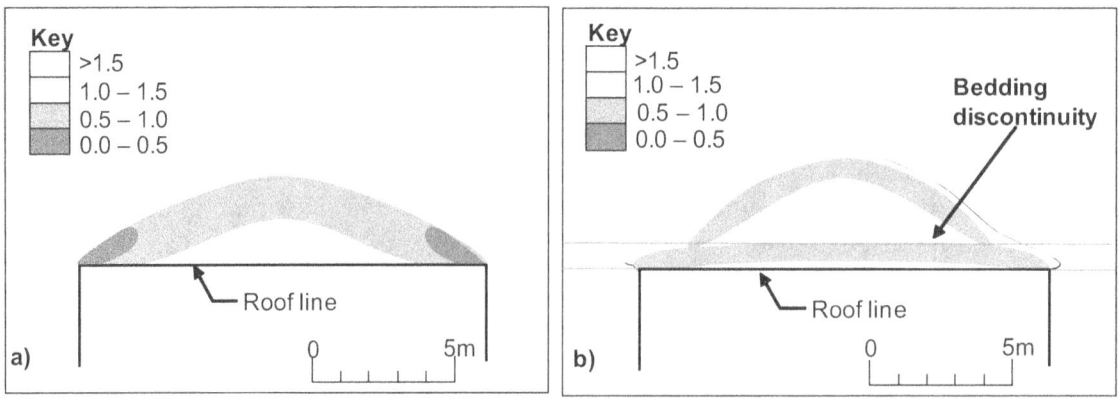

Figure 36. Vertical cross section through a heading showing rock failure index values (a) without bedding discontinuities and (b) with a bedding discontinuity 1 m (3.3 ft) above the roofline.

A third model was set up in which three bedding discontinuities 1 m (3.3 ft) apart were introduced above the roofline, as shown in Figure 37a. The potential failure now extends 5 m (17 ft) above the roofline as beam deflection and stress redistribution continues further into the roof.

In the final case the roof is modeled as a thinly bedded rock using model elements that assume that each element in the model contains multiple horizontal planes of weakness that can shear. The strength of these ubiquitous weaknesses was set equal to that of the bedding discontinuities described above. The stability index results, in Figure 37b, show that the extent of potential failure is much greater, now extending about 7 m (23 ft) above the roofline. Inspection of the results show that slip along the roof beds allowed more roof deflection to occur, which reduced the confinement in the roof.

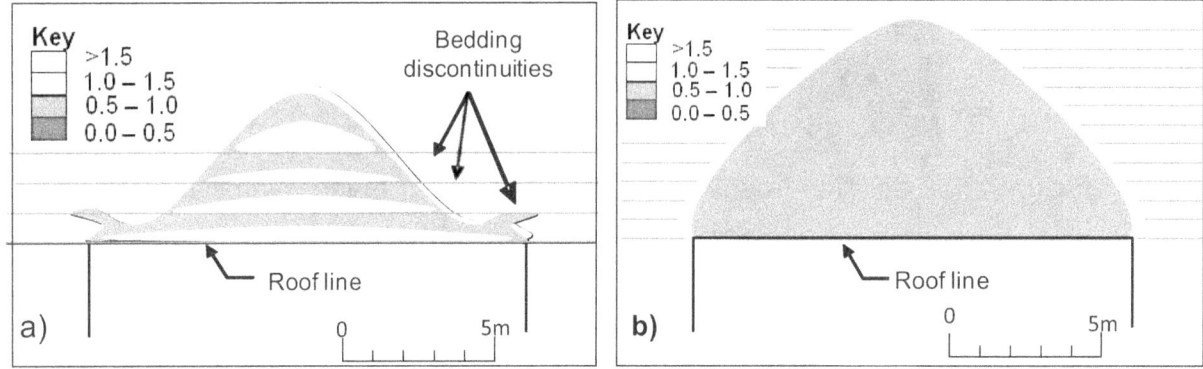

Figure 37. Vertical cross section through a heading showing rock failure index values (a) with three 1-m thick (3.3 ft) bedding discontinuities in the roof and (b) thinly laminated roof.

These numerical model results show that elevated horizontal stress can cause extension-type failure in the roof of stone mine excavations in the absence of weak bedding planes. The presence of bedding planes exacerbates the situation by shifting the stress higher into the roof, which results in a greater height of potential roof failure.

Once a stress-induced roof fall has occurred, it can be costly and difficult to arrest the lateral extension of the fall into adjacent areas. Avoidance of these falls through layout modifications has proven to be very successful in several operating mines [Iannacchione et al. 2003]. First, the direction of the major horizontal stress must be established, which can be determined by various stress measurement techniques or can be inferred from stress-related roof failures [Mark and Mucho 1994]. The layout is then modified so that the main development direction is parallel to the maximum horizontal stress and the amount of unfavorably oriented crosscut development is minimized [Parker 1973]. A further modification that has proven to be successful is offsetting the crosscuts and increasing the length of the pillars, so that a continuous path does not exist along which a roof fall can progress through the layout. Offset crosscuts also result in three-way intersections, which are more stable than the four-way intersections. Modifying a layout in this manner will not necessarily eradicate all stress-related problems, but has been shown to considerably reduce these problems [Kuhnhein and Ramer 2004].

Figure 38 shows a mine layout that has been optimized for horizontal stress. The main heading direction is parallel to the maximum horizontal stress; pillars are elongated so that unfavorably oriented crosscuts are minimized. The crosscuts are narrower than the headings and are offset so that potential stress-related roof falls will abut against solid pillar ribs, rather than snake through the layout.

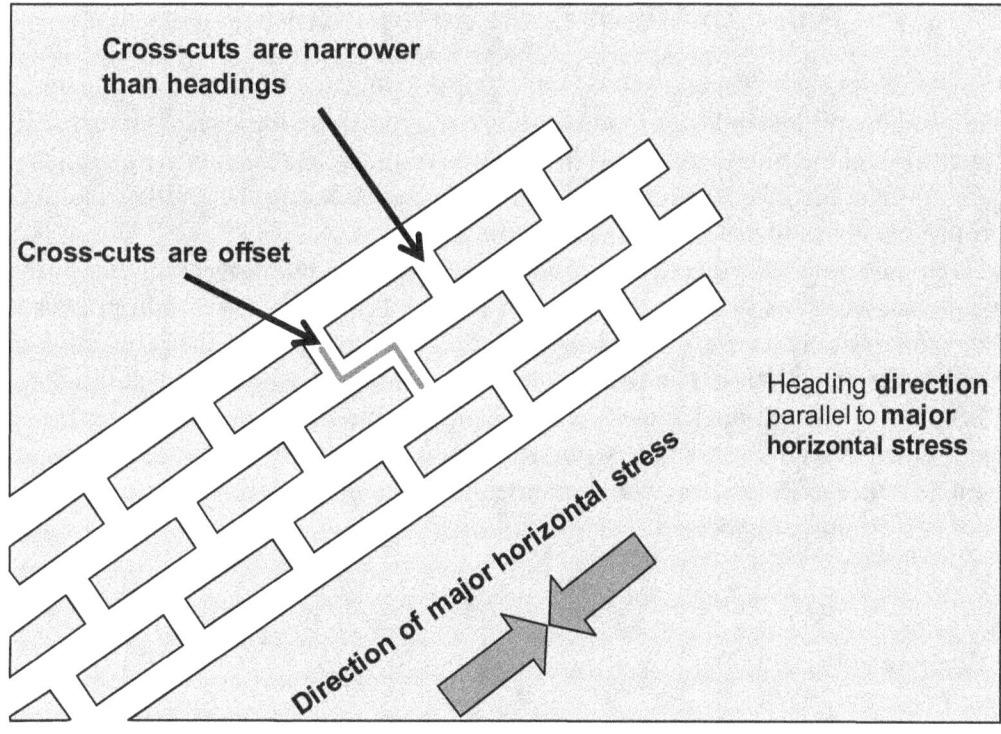

Figure 38. Diagram showing room-and-pillar layout modified to minimize the potential impact of horizontal, stress-related damage.

Roof Support

Roof reinforcement in the relatively strong-bedded rock encountered in stone mines can have one or more objectives. Depending on the geological conditions, the support system can be expected to:

- Provide suspension support for a potentially unstable roof beam.
- Provide local support to potentially unstable blocks in the roof.
- Combine thinly laminated roof into a thicker, stronger unit.
- Provide surface control when progressive spalling and small rock falls occur.

The above support functions can usually be achieved by the 1.8-m (6-ft) and 2.4-m (8-ft) bolts used in the stone mines. When poor ground is encountered locally or when horizontal stress-related roof failures occur, supplementary bolting, steel straps and screen, and longer cable bolts have been used with mixed success to halt the lateral extension of these large roof falls.

From a design point of view, a stone mine is unlikely to be economically feasible if heavy support such as cable bolts and screen would be required on a daily basis. Such rock conditions would probably require reduced excavation spans, and the support costs would be prohibitive. The first objective in designing an underground stone mine should be to confirm that the rock mass quality is adequate for creating the typical 13 m (43 ft) roof spans without resorting to elaborate support systems.

Pillar and Roof Span Design Guidelines

Designing stable pillars and roof spans for underground stone mines is an integrated process. The roof spans affect the pillar stress, and the pillar layout can have a significant impact on roof stability. The design, therefore, should be conducted by considering the ability of both the pillars and the roof spans to produce a stable overall mine layout.

The design guidelines listed below start with developing a clear understanding of the geotechnical characteristics of the rock mass that will be mined. Next, an appropriate roof span and roof horizon must be selected. The main development direction for the production areas should then be determined, based on rock structures and the likely horizontal stress direction. Support needs are addressed next. Once the roof design components are complete, the dimensions of the pillars are set, and any possible changes to the pillar layout for horizontal stress are made. Once a design has been implemented, the pillars and roof are checked to verify that they are performing as expected.

Geotechnical Characterization

Designing stable pillars and roof spans for stone mines can be successfully conducted if adequate geotechnical investigations are conducted before the design phase. Such investigations are best conducted by experienced ground control specialists and are likely to include rock strength testing, core logging, bedding layering assessment, joint orientation assessment, and rock mass classification. If horizontal, stress-related issues are expected, stress measurements can assist in providing an indication of the orientation and magnitude of the maximum horizontal stress.

Useful information can be obtained from nearby mines that are operating under similar conditions. A particularly useful piece of information would be to identify whether horizontal, stress-related roof problems exist and to know the orientation of the stress-related damage. This information can go a long way in selecting the orientation of the main headings in the proposed mine.

The geotechnical data should be used to confirm that the rock conditions are similar to those observed in the stone mines that were included in this study. The RMR [Bieniawski 1989] should exceed a value of 60.0 and the UCS of the rock should exceed 45 MPa (6,400 psi). The absence of weak, softening bands within the mining horizon should be confirmed. These weak bands can have a significant detrimental effect on pillar strength. Similarly, the presence or absence of large, angular discontinuities should be identified because they should be accounted for in the pillar strength determination.

During initial scoping studies, the geotechnical data, such as the rock strength, rock mass rating, and presence of angular discontinuities may be unknown. In these cases, conservative estimates should be used; however, the appropriate site-specific geotechnical data must be obtained for the final design.

Roof Span Selection

Past experience has shown that stable roof spans in the range of 10 m to 15 m (33 ft to 50 ft) have been regularly achieved in underground stone mines. NIOSH studies have shown little correlation between mining roof spans and rock quality, mainly because there is such a small range of rock qualities in operating mines.

For an initial design it might be prudent to design for no more than 12 m (40 ft) spans; and, if rock conditions and monitoring of actual roof performance warrants it, the spans can be increased incrementally. There is limited experience with spans that are greater than 15 m (50 ft).

The need for roof support is strongly related to the thickness of the first rock bed in the roof of the excavations. Modifying the roof span within the 10 to 15 m (33 to 50 ft) range will not necessarily change the need for rock support.

Selecting the Roof Horizon

The location of the roofline relative to pronounced bedding planes or lithology changes should be identified next. Experience has shown that if the immediate roof beam is less than 1.2 m (4 ft) thick, it is highly likely that it will be unstable. Thicker roof beams may be required if excessive horizontal stresses are encountered. Mines in the Appalachian Highlands region, where horizontal stress problems exist, tend to have roof beams that are in the range of 2.7 to 5 m (9 to 16 ft).

Persistent parting planes can be selected to form the roofline if they are present at a convenient location in the formation being mined. Using a preexisting parting plane as the roofline helps to act as a marker and usually provides a clean breaking surface for blasting operations. Many of the mines that do not use roof supports have a natural parting as the roofline.

Orientation of Headings

The direction of the headings in the production areas should be favorably oriented to any expected horizontal stress and the prevalent jointing. As with any underground excavation layout, it is preferable to intersect the main joint strike direction as near to perpendicular as possible. Because room-and-pillar mines have two orthogonal directions of mining, the heading direction should be favored over the crosscut direction when selecting the orientation of the layout.

If the orientation of the maximum horizontal field stress is known, and stress-related problems are anticipated, the heading direction should be oriented parallel to the direction of major horizontal stress, with due consideration of joint orientations and crosscut stability. It is often a compromise to select the final heading orientation. Modifications can also be made to the pillar layout to enhance roof stability in high horizontal stress conditions. These modifications were summarized in Figure 38.

Roof Support Considerations

Depending on the characteristics of the immediate roof, basic support in the form of patterned rock bolts may be required. The importance of the thickness of the first beam in the roof, the orientation of excavations relative to the maximum horizontal stress, and characteristics of rock joints will determine whether and how much support is required. Rock bolts in the range of 1.8 to 2.4 m (6 to 8 ft) are most commonly used in stone mines. Mines that do not use bolting are located in formations with a favorable combination of geological conditions, and they conduct blasting practices that maintain an unbroken roof horizon.

Pillar Design

Pillar design can be conducted using equation 7 provided the rock mass quality, mining dimensions, and depth of cover remain within similar bounds as those that were used to develop the equation.

Should weak bands that may extrude from within pillars be identified during the geotechnical assessment, equation 7 should not be used. Weak-banded pillars were specifically excluded from the study. In such a case, the advice of a rock engineering specialist should be sought. Similarly, further investigation by rock engineering specialists will be required if more than about 30% of the pillars in a layout are expected to be intersected by large, angular discontinuities that dip from 30° to 70°.

Pillars having a width-to-height ratio of less than 0.8 should be avoided. Slender pillars are highly susceptible to the impact of large, angular discontinuities and are inherently weaker than wider pillars because the pillar core is unconfined. The potential for extension-type fractures to propagate right through these slender pillars is another reason for avoiding them.

Pillars that are designed using equation 7 should have a factor of safety of at least 1.8, which represents the lower bound of current experience. The shaded area in Figure 25 shows the recommended area for pillar design. This chart shows that there are only two cases of stable pillar layouts below a factor of safety of 1.8, and many of the failed cases plot below this value. A lower bound factor of safety of 1.8 is, therefore, recommended.

Pillars should be designed so that the average pillar stress does not exceed 25% of the UCS, which is within the limits of past experience. The presence of extension-type fractures within stone pillars that are loaded to high stress levels can have unexpected effects on their strength. Detailed investigation by rock engineering specialists coupled with systematic monitoring of pillar performance is recommended if the pillar stress is expected to exceed 25% of the UCS of the rock.

The shaded area in Figure 25 shows the recommended area for pillar design based on the outcomes of this research. Designs that fall outside the shaded area have an elevated risk of instability and require further investigation by rock engineering specialists.

Layout Modification for Horizontal Stress

A simple, square pillar layout, with headings and crosscuts of equal width, is sufficient in most cases. However, if horizontal stress-related instability is expected, the pillar layout can be modified to improve the likelihood of success. Possible layout modifications are shown in Figure 38, which include: orienting the main development direction parallel to the maximum horizontal stress, offsetting crosscuts to arrest the lateral expansion of stress-related falls, and increasing the length of pillars so that the number of unfavorably oriented cross-cuts is reduced.

Monitoring and Verification

Once the roof span and pillar design has been finalized and mining is underway, monitoring should be implemented to verify the stability of the roof and pillars. Monitoring results can be used to identify potential stability problems before they occur and may indicate that a change in the design is required. Monitoring technologies that are available include borehole-video logging [Ellenberger 2009], roof deflection monitoring [Marshall et al. 2000], roof stability mapping using the Roof Fall Risk Index (RFRI), [Iannacchione et al. 2006] and microseismic monitoring of rock fracture [Iannacchione et al. 2004; Ellenberger and Bajpayee 2007].

Summary and Conclusions

A study of pillar and roof span performance in stone mines that are located in the Eastern and Midwestern United States showed that various stability issues can be addressed by appropriate pillar and roof span design. Pillars can be impacted by rock joints, large angular discontinuities and can exhibit rib spalling at elevated stresses. Thin weak beds in the pillars, although rare, can have a significant impact by reducing pillar strength. If the roof strata are bedded, beam deflection and buckling can result in roof failure. The roof can also be impacted by large discontinuities and the effects of horizontal stress.

A pillar design procedure is proposed that takes into consideration the rock strength, pillar dimensions and the potential impact of large angular discontinuities. Based on the proposed pillar design procedure and the observed performance of pillars in stone mines, a safety factor of at least 1.8 is recommended for pillar design. A lower limit pillar width-to-height ratio of 0.8 is also recommended. Designs that fall outside these limits have an elevated risk of instability and further investigation by rock engineering specialists is required.

A roof span design procedure is also proposed that systematically addresses each of the main stability issues. The procedure focuses on selecting an appropriate mining horizon and mining direction. The importance of the thickness of the first bed in the roof and the likelihood for added rock bolting is described. Layout modifications are described that can be made to reduce the incidence of horizontal-stress-related instability.

Both the pillar design and roof span guidelines require that a good understanding be obtained of the geotechnical characteristics of the formation being mined. The essential data are the uniaxial compressive strength of the rock, characteristics of the discontinuities and the rock mass classification. Knowledge of the magnitude and orientation of the stress field can assist in orienting the layout appropriately.

The design procedures are based on observation of the actual performance of pillars and roof spans in stone mines within the Eastern and Midwestern United States. The guidelines should only be used for design under similar geotechnical conditions.

References

Barton N, Lien R, Lunde J [1974]. Engineering classification of rock masses for the design of tunnel support. Rock Mech. Rock Eng. 6(4):189–236.

Bieniawski ZT [1989]. Engineering rock mass classifications: a complete manual for engineers and geologists in mining, civil and petroleum engineering. New York, NY: John Wiley and Sons, Inc.

Brady BHG, Brown ET [1985]. Rock mechanics for underground mining. London, England: George Allen and Unwin.

Brann RW, Freas RC [2003]. Multiple level room and pillar mining in limestone. SME preprint 03-058. Littleton, CO: Society for Mining, Metallurgy, and Exploration, Inc.

Carmack J, Dunn B, Flach M, Sutton G [2001]. The viburnum trend underground. underground mining methods: engineering fundamentals and international case studies. Hustrulid WA, Bullock RC, eds.. Society for Mining Metallurgy and Exploration, pp. 89–94.

Diederichs MS, Coulson A, Falmagne V, Rizkalla N, Simser B [2002]. Application of rock damage limits to pillar analysis at Brunswick Mine. In: Hammah et al., eds. Proceedings of NARMS-TAC 2002, Toronto, Ontario, Canada: University of Toronto, pp. 1325–1332.

Diederichs MS [2002]. Stress induced damage accumulation and implications for hard rock engineering. In Hammah et al., eds. Proceedings of NARMS-TAC 2002, Toronto, Ontario, Canada: University of Toronto, pp. 3–12.

Dolinar DR [2003]. Variation of horizontal stresses and strains in mines in bedded deposits in the eastern and midwestern United States. In: Peng SS, Mark C, Khair AW, Heasley KA, eds. Proceedings of the 22nd International Conference on Ground Control in Mining. Morgantown, WV: West Virginia University, pp. 178–185.

Dolinar DR, Esterhuizen GS [2007]. Evaluation of the effects of length on strength of slender pillars in limestone mines using numerical modeling. In: Peng SS, Mark C, Finfinger GL, Tadolini SC, Khair AW, Heasley KA, Luo Y, eds. Proceedings of the 26th International Conference on Ground Control in Mining. Morgantown, WV: West Virginia University, pp. 304–313.

Ellenberger JL [2009]. A roof quality index for stone mines using borescope logging. In: Peng SS, Barczak TM, Mark C, Tadolini SC, Finfinger GL, Heasley KA, Luo Y, eds. Proceedings of the 28th International Conference on Ground Control in Mining. Morgantown, WV: West Virginia University, pp. 143–148.

Ellenberger JL, Bajpayee TS [2007]. An evaluation of microseismic activity associated with major roof falls in a limestone mine: a case study. SME preprint 07-103. Littleton, CO: Society for Mining, Metallurgy, and Exploration, Inc.

Esterhuizen GS [2000]. Jointing effects on pillar strength. In: Peng SS, Mark C, eds. Proceedings of the 19th International Conference on Ground Control in Mining. Morgantown, WV: West Virginia University, pp. 286–290.

Esterhuizen GS [2006]. An evaluation of the strength of slender pillars. In: Yernberg WR, ed. Transactions of Society for Mining, Metallurgy, and Exploration, Inc. Vol. 320. Littleton, CO: Society for Mining, Metallurgy, and Exploration, Inc., pp. 69–76.

Esterhuizen GS, Ellenberger JL [2007]. Effects of weak bands on pillar stability in stone mines: field observations and numerical model assessment. In: Peng SS, Mark C, Finfinger GL, Tadolini SC, Khair AW, Heasley KA, Luo Y, eds. Proceedings of the 26th International Conference on Ground Control in Mining. Morgantown, WV: West Virginia University, pp. 320–326.

Esterhuizen GS, Dolinar DR, Ellenberger JL, Prosser LJ Jr., Iannacchione AT [2007]. Roof stability issues in underground limestone mines in the United States. In: Peng SS, Mark C, Finfinger GL, Tadolini SC, Khair AW, Heasley KA, Luo Y, eds. Proceedings of the 26th International Conference on Ground Control in Mining. Morgantown, WV: West Virginia University, pp. 336–343.

Esterhuizen GS, Dolinar DR, Ellenberger JL [2008]. Pillar strength and design methodology for stone mines. In: Peng SS, Tadolini SC, Mark C, Finfinger GL, Heasley KA, Khair AW, Luo Y, eds. Proceedings of the 27th International Conference on Ground Control in Mining. Morgantown, WV: West Virginia University, pp. 241–253.

Gale WJ [1999]. Experience of field measurement and computer simulation methods of pillar design. In: Mark C, Heasley KA, Iannacchione AT, Tuchman RJ, eds. Proceedings of the Second International Workshop on Coal Pillar Mechanics and Design. Pittsburgh, PA: U.S. Department of Health and Human Services, Public Health Service, Centers for Disease Control and Prevention, National Institute for Occupational Safety and Health, DHHS (NIOSH) Publication No. 99-114, IC 9448, pp. 49–61.

Galvin JM, Hebblewhite BK, Salamon MDG [1999]. University of New South Wales coal pillar strength determinations for Australian and South African mining conditions. In: Mark C, Heasley KA, Iannacchione AT, Tuchman RJ, eds. Proceedings of the Second International Workshop on Coal Pillar Mechanics and Design. Pittsburgh, PA: U.S. Department of Health and Human Services, Public Health Service, Centers for Disease Control and Prevention, National Institute for Occupational Safety and Health, DHHS (NIOSH) Publication No. 99-114, IC 9448, pp. 63–71.

Grau RH III, Krog RB, Robertson SB [2006]. Maximizing the ventilation of large-opening mines. In: Mutmansky JM, Ramani RV, eds. Proceedings of the 11th U.S./North American Mine Ventilation Symposium (University Park, PA, June 5-7, 2006). London, England: Taylor & Francis Group, pp. 53–59.

Hajiabdolmajid V, Martin CD, Kaiser PK [2000]. Modelling brittle failure of rock. In: Girard J, Liebman M, Breeds C, Doe T, eds. Pacific Rocks 2000. Proceedings of the Fourth North

American Rock Mechanics Symposium. Rotterdam, Netherlands: A.A. Balkema Publishers, pp. 991–998.

Harr ME [1987]. Reliability based design in civil engineering. New York, NY: McGraw-Hill.

Heasley KA, Agioutantis Z [2001]. LAMODEL—a boundary element program for coal mine design. In: Proceedings of the 10th International Conference on Computer Methods and Advances in Geomechanics. Tucson, Arizona, pp. 9–12.

Hoek E, Kaiser PK, Bawden WF [1995]. Support of underground excavations in hard rock. Rotterdam, Netherlands: A.A. Balkema.

Hoek E, Carranza-Torres C, Diederichs MS, Corkum B [2008]. Integration of geotechnical and structural design in tunnelling. In: Proceedings of the University of Minnesota 56th Annual Geotechnical Engineering Conference. Minneapolis, MN, pp. 1–53.

Hustrulid WA [1976] A review of coal pillar strength formulas. Rock Mech. and Rock Eng. *8*(2):115–145.

Hutchinson DJ, Diederichs MS [1996]. Cablebolting in underground mines. Bitech Publishers Ltd.: Canada.

Iannacchione AT [1999]. Analysis of pillar design practices and techniques for U.S. limestone mines. Trans. Inst. Min. Metall. (sect. A: Min. Industry), September–December, *108*:A152–A160.

Iannacchione AT, Coyle PR [2002]. An examination of the Loyalhanna limestones structural features and their impact on mining and ground control practices. In: Peng SS, Mark C, Khair AW, Heasley KA, eds. Proceedings of the 21st International Conference on Ground Control in Mining. Morgantown, WV: West Virginia University, pp. 218–227.

Iannacchione AT, Dolinar DR, Prosser LJ Jr., Marshall TE, Oyler DC, Compton CS [1998]. Controlling roof beam failures from high horizontal stresses in underground stone mines. In: Peng SS, ed. Proceedings of the 17th International Conference on Ground Control in Mining. Morgantown, WV: University of West Virginia, pp. 102–112.

Iannacchione AT, Dolinar DR, Mucho TP [2002]. High-stress mining under shallow overburden in underground U.S. stone mines. In: Proceedings of the First International Seminar on Deep and High-Stress Mining. Nedlands, Australia: Australian Centre for Geomechanics, section 32, pp. 111.

Iannacchione AT, Marshall TE, Burke L, Melville R, Litsenberger J [2003]. Safer mine layouts for underground stone mines subjected to excessive levels of horizontal stress. Min Eng *55*(4):25–31.

Iannacchione AT, Batchler TJ, Marshall TE [2004]. Mapping hazards with microseismic technology to anticipate roof falls: a case study. In: Peng SS, Mark C, Finfinger GL, Tadolini

SC, Heasley KA, Khair AW, eds. Proceedings of the 23rd International Conference on Ground Control in Mining. Morgantown, WV: West Virginia University, pp. 327–333.

Iannacchione AT, Esterhuizen GS, Schilling S, Goodwin T [2006]. Field verification of the roof fall risk index: a method to assess strata conditions. In: Peng SS, Mark C, Finfinger GL, Tadolini SC, Khair AW, Heasley KA, Luo Y, eds. Proceedings of the 25th International Conference on Ground Control in Mining. Morgantown, WV: West Virginia University, pp. 128–137.

Kaiser PK, Diederichs MS, Martin DC, Steiner W [2000]. Underground works in hard rock tunneling and mining. Keynote Lecture, Geoeng2000, Melbourne, Australia: Technomic Publishing Co., pp. 841–926.

Krauland N, Soder PE [1987]. Determining pillar strength from pillar failure observations. Eng Min J *8*:34–40.

Kuhnhein G, Ramer R [2004]. The influence of horizontal stress on pillar design and mine layout at two underground limestone mines. In: Peng SS, Mark C, Finfinger GL, Tadolini SC, Heasley KA, Khair AW, eds. Proceedings of the 23rd International Conference on Ground Control in Mining. Morgantown, WV: West Virginia University, pp. 311–319.

Iannacchione AT, Batchler TJ, Marshall TE [2004]. Mapping hazards with microseismic technology to anticipate roof falls: a case study. In: Peng SS, Mark C, Finfinger GL, Tadolini SC, Heasley KA, Khair AW, eds. Proceedings of the 23rd International Conference on Ground Control in Mining. Morgantown, WV: West Virginia University, pp. 327–333.

Lane WL, Yanske TR, Roberts DP [1999]. Pillar extraction and rock mechanics at the Doe Run Company in Missouri 1991 to 1999. In: Amadei B, Kranz RL, Scott GA, Smeallie PH, eds. Proceedings of the 37th US Rock Mechanics Symposium. Rotterdam, Netherlands: A.A. Balkema Publishers, pp. 285–292.

Laubscher DH [1990]. A geomechanics classification system for the rating of rock mass in mine design. J S Afr. Inst. Min. Metall. *90*(10):257–273.

Lunder PJ [1994]. Hard rock pillar strength estimation—an applied approach [M.S. Thesis]. Vancouver, BC: University of British Columbia, Department of Mining and Mineral Process Engineering.

Mark C [1999]. Empirical methods for coal pillar design. In: Mark C, Tuchman RJ, eds. Proceedings: New technology for ground control in retreat mining. Cincinnati, OH: U.S. Department of Health and Human Services, Public Health Service, Centers for Disease Control and Prevention, National Institute for Occupational Safety and Health, DHHS (NIOSH) Publication No. 2000-151, pp. 145–154.

Mark C, Mucho TP [1994]. Longwall mine design for control of horizontal stress. In: Mark C, Tuchman RJ, Repsher RC, Simon CL, eds. New Technology for Longwall Ground Control. Proceedings: U.S. Bureau of Mines Technology Transfer Seminar. Pittsburgh, PA: U.S. Department of the Interior, Bureau of Mines, SP 01–94, pp. 53–76.

Marshall TE, Prosser LJ Jr., Iannacchione AT, Dunn M [2000]. Roof monitoring in limestone mines: experience with the roof monitoring safety system (RMSS). In: Peng SS, Mark C, eds. Proceedings of the 19th International Conference on Ground Control in Mining. Morgantown, WV: West Virginia University, pp. 185–191.

Martin CD, Chandler N [1994]. The progressive fracture of Lac du Bonnet granite. Int J Rock Mech and Min Sci *31*(6):643–659.

Martin CD, Maybee WG [2000]. The strength of hard rock pillars. Int J Rock Mech and Min Sci *37*:1239–1246.

Mathews KE, Hoek DC, Wyllie DC, Stewart SBV [1980]. Prediction of stable excavation spans for mining at depths below 1,000 metres in hard rock. Report to Canada Centre for Mining and Energy Technology (CANMET), Department of Energy and Resources; DSS File No. 17SQ.23440-0-90210. Ottawa, Canada.

MSHA [2009]. Statistics Single Source Page, [htpp://www.msha.gov/stats/statistics.htm]. Date accessed: October 2009.

Nickson SD [1992]. Cable support guidelines for underground hard rock mine operations [M.S. Thesis]. Vancouver, B.C.: University of British Columbia, Dept. Mining and Mineral Process Engineering.

Parker J [1973]. How to design better mine openings: practical rock mechanics for miners. Eng Min *174*(12):76–80.

Potvin Y [1988]. Empirical open stope design in Canada [Ph.D. Dissertation]. Vancouver, B.C.: University of British Columbia, Dept. Mining and Mineral Process Engineering.

Pritchard CJ, Hedley DGF [1993]. Progressive pillar failure and rockbursting at Denison Mine. In: Young R.P., ed. 3rd International Symposium on Rockbursts and Seismicity in Mines, Queen's University, Kingston, ON, August 1993. Rotterdam, Netherlands: A.A. Balkema,, pp. 111–116.

Roberts D [2005]. Golder Associates [Personal communication].

Roberts D, Tolfree D, McIntyre H [2007]. Using confinement as a means to estimate pillar strength in a room and pillar mine. In: Eberhardt E, Stead D, Morrison T, eds. Proceedings of the First Canada-U.S. Rock Mechanics Symposium (Vancouver, British Columbia, Canada, May 27-31, 2007). Taylor & Francis Ltd., *2*:1455–1461.

Salamon MDG [1970]. Stability, instability and the design of pillar workings. Int J Rock Mech Min Sci & Geomech Abstr *7*(6):613–631.

Salamon MDG, Munro AH [1967]. A study of the strength of coal pillars. J South Afr. Inst. Min. Metall. *68*:55–67.

Salamon MDG, Canbulat I, Ryder JA [2006]. Seam-specific pillar strength formulae for south african collieries. In: Yale DP, Holtz SC, Breeds C, Ozbay U, eds. Proceedings of the 41st U.S. Rock Mechanics Symposium (Golden, CO, June 17–21, 2006). Alexandria, VA: American Rock Mechanics Association, Paper 06-1154.

Stacey TR [1981]. A simple extension strain criterion for fracture of brittle rock. Int J Rock Mech Min Sci Geomechan Abstr *18*:469–474.

Stace TR, Yathavan K [2003]. Examples of fracturing of rock at very low stress levels. In: Proceedings of the ISRM 2003 Congress, Technology Roadmap for Rock Mechanics. S Afr Inst Min Metall, pp. 1155–1159.

U.S. Geological Survey [2009]. Paleontology web page [http://geology.er.usgs.gov/paleo]. Date accessed: October 2009.

Wagner, H. [1992]. Pillar design in South African collieries. In: Iannacchione AT, Mark C, Repsher RC, Tuchman RJ, Jones CC, eds. Proceedings of the Workshop on Coal Pillar Mechanics and Design. U.S. Department of the Interior, Bureau of Mines, IC 9315, pp. 283–301.

World Stress Map Project. [2009]. http://dc-app3-14.gfz-potsdam.de /pub /introduction/ introduction_frame.html. Date accessed: October 2009.

Zipf RK Jr [2001]. Pillar design to prevent collapse of room-and-pillar mines in underground mining methods: engineering fundamentals and international case studies. Hustrulid WA, Bullock RC, eds. Society for Mining, Metallurgy and Exploration, pp. 493–511.

Zipf RK [2008]. National institute for Occupational Safety and Health [Personal Communication].

www.ingramcontent.com/pod-product-compliance
Lightning Source LLC
Chambersburg PA
CBHW081847170526
45167CB00007B/2923